ABSTRACT SETS
and
FINITE ORDINALS

An Introduction to
the Study of Set Theory

G. B. KEENE

DOVER PUBLICATIONS, INC.
Mineola, New York

To Mary

Bibliographical Note

This Dover edition, first published in 2007, is an unabridged republication of the work originally published by Pergamon Press, New York, in 1961.

Library of Congress Cataloging-in-Publication Data

Keene, G. B. (Geoffrey Bourton)
  Abstract sets and finite ordinals : an introduction to the study of set theory / G.B. Keene.
    p. cm.
  Originally published: Oxford ; New York : Pergamon Press, 1961.
  Includes index.
  ISBN-13: 978-0-486-46249-3
  ISBN-10: 0-486-46249-8
  1. Set theory. 2. Logic, Symbolic and mathematical. I. Title.

QA248.K4 2007
511.3'22—dc22

2007023371

Manufactured in the United States of America
Dover Publications, Inc., 31 East 2nd Street, Mineola, N.Y. 11501

CONTENTS

PREFACE

IN this book an attempt is made to present, in a reasonably simple way, the outlines of a relatively complex subject. Although there are now many textbooks devoted to the relationship between mathematics and the logic of classes, very few of them are of much help to the newcomer. They usually presuppose either some knowledge of mathematical logic, despite the fact that few mathematics students are trained in logic, or else considerable competence in mathematics, thus neglecting the needs of many philosophy students. The present text aims to meet the needs of students both of mathematics and of philosophy, as well as those of the general reader. For this reason a preliminary survey is made, in Part I, of the few symbols of logic used in the second half of the text. It can safely be taken as read by anyone who has studied elementary modern logic. At the same time, no knowledge of mathematics is presupposed.

The basis of Part II is the System of Axiomatic Set Theory of Paul Bernays, originally published as a series of articles in the *Journal of Symbolic Logic*[1]. In these articles Bernays makes as little explicit use as possible of the formalism of logic. But in doing so he makes implicit use of the professional reader's background knowledge. Consequently much of the intricate reasoning which lies just below the surface, so to speak, is beyond the grasp of readers new to the subject. The purpose of this book is to present a fragment of the Bernays Theory in a version which, while it makes explicit use of a certain amount of formalism, calls for no previous acquaintance with the subject.

The result, it is hoped, will serve as an introduction both to the complete Bernays theory and to the techniques of set theory in general. It should also provide some point of application for

[1] See References, p. 103; a fully formalized version is now available, in *Axiomatic Set Theory* by Bernays and Fraenkel, which appeared after the present book was completed.

elementary logic, less trivial than the usual testing-for-validity
of artificially devised arguments.

I am indebted to Professor Bernays for his criticisms of the
first draft of the book. His suggestions as well as his kindly
interest in it considerably lightened the burden of final pre-
paration and correction of the typescript. I am also indebted
to Mr. P. Geach of Birmingham, who read the typescript in an
early form, to Professor D. J. O'Connor of Exeter University
and Professor I. N. Sneddon of Glasgow University for their
help and encouragement, and to the Editor of the *Journal of
Symbolic Logic*, Professor S. C. Kleene, for his approval of my
adaptation of material published in that Journal.

Exeter   G. B. K.

PART I
THE ELEMENTS OF SET THEORY

THE BASIC LOGICAL CONCEPTS

1.1. Introduction

THE symbolism of elementary logic is both simple and efficient. By its means elegant and powerful logical calculi can be set up for the purpose of formalizing a scientific theory. But familiarity with the techniques involved in setting up these logical calculi is not essential to an understanding of the present text. All that is required is an understanding of the meaning and interrelation of the following symbols:

$$\sim, \quad \cdot, \quad \lor, \quad \supset, \quad \equiv, \quad (x)\phi, \quad (\exists x)\phi, \quad \varepsilon$$

All of them translate straightforwardly into a familiar word or phrase of ordinary speech. But they are not mere shorthand symbols. For each differs from its counterpart in ordinary speech in virtue of the fact that it is precisely defined, whereas the words of ordinary speech are not. We shall formulate, explain and illustrate the definition of each in turn. In doing so we shall make use of the two groups of letters: x, y, z, w, and: P, Q, R, S, which we call **variables**. They are variables in the sense that they keep a place open in a formula, just as the x in x^2 keeps a place open for any number we may wish to put there. The type of entity that may be put in place of our variables will be called the **range of values** of those variables. In fact the range of values of the variables x, y, z, w, is any individual object of any kind, and the range of values of the variables P, Q, R, S, is any proposition. Finally we shall use the notation **T**, to mean any true proposition, and **F** to mean any false proposition, and we shall refer to the symbols listed above as **logical constants**.

1.2. The Logical Constants

1.21. *"Member of"*

When the Greek letter ε occurs between two variables as in:

$$x \, \varepsilon \, y$$

the range of values of the variable y is, thereby, confined to classes. Thus if the value *John Doe* is given to the variable x and the value *mortals* to the variable y, the result is the formula:

$$\text{John Doe } \varepsilon \text{ mortals}$$

which may be read:

"John Doe is a member of the class Mortals"

Again if we write Nn as the name of the class Natural Numbers, then the formula:

(I) $\qquad\qquad\qquad x \, \varepsilon \, \text{Nn}$

gives rise to the following formulae, each of which represents a true statement:

(II) $\qquad\qquad\qquad 0 \, \varepsilon \, \text{Nn}$

$\qquad\qquad\qquad\qquad 1 \, \varepsilon \, \text{Nn}$

$\qquad\qquad\qquad\qquad 2 \, \varepsilon \, \text{Nn}$

$\qquad\qquad\qquad\qquad \ldots \text{etc.}$

Any formula of the type (I) above will be called a **propositional function,** where this term is intended to mean: an expression which becomes a proposition (true or false) when values are assigned (suitably) to all the variables (of the sort: x, y, z) occurring in it. The propositional function is the most important among the various types of formula with which we shall be concerned in Part II. The word **formula,** however, will be used to refer quite generally to: either a propositional variable (e.g. P, Q etc.) standing alone, or a propositional function, or any combination of these properly constructed by means of the logical constants or, finally, a symbolized proposition (e.g. (II) above). In this connection we shall, for convenience, use the phrase "formula which represents a true (false) proposition" in the following way. A formula may represent a true (false) proposition either directly (as in the case of a symbolized proposition), or indirectly (as in the case of a formula containing variables which is assumed, for the sake of a particular argument, to give rise to some particular but unspecified true (false) proposition, by a suitable assignment of values to the variables).

1.22. "*Not*"

We use the sign "~" (read "tilde") *immediately to the left* of a formula. It is defined by a matrix which shows whether a formula having this sign as its first symbol, represents a true or a false proposition. The matrix shows this by reference to the formula to which the sign is adjoined:

P	$\sim P$
T	F
F	T

Thus if the sign occurs against a formula representing a true proposition, the result is a formula representing a false proposition (and vice versa); so that if it occurs against a *propositional variable P* which has a true proposition as its intended value, the result will be a formula representing a false proposition (and vice versa). Similarly, if it occurs against a *propositional function* having a true proposition as its intended value, the result will be a formula representing a false proposition (and vice versa). For example, if G is used to name the class of Greeks, then the formula:

$$x \; \varepsilon \; G$$

gives rise to the false proposition:

(III) $Mz \; \varepsilon \; G$

if we are using Mz to name, say, Mozart. Therefore, by the above matrix, if we substitute (III) for P in the formula $\sim P$, the result is a formula representing a true proposition, namely:

$$\sim(Mz \; \varepsilon \; G)$$

(i.e. "Mozart is not a Greek")

Again, if we substitute $\sim(0 \; \varepsilon \; Nn)$ for P in the formula $\sim P$, the result is the following true symbolized proposition:

$$\sim(\sim(0 \; \varepsilon \; Nn))$$

(i.e. "0 is not not a member of the class Natural Numbers").

1.23. *"And"*

We use the sign · (read "dot") *between* formulae, and it is defined by the following matrix:

P	Q	$P \cdot Q$
T	T	T
F	T	F
T	F	F
F	F	F

The matrix shows that if dot occurs between two formulae each of which represents a true proposition, the result is a formula representing a true proposition. In every other possible case the result is shown to be a formula representing a false proposition. For example, using the same name symbols as before, if we substitute (0 ε Nn) for P, and (Mz ε G) for Q in the formula $P \cdot Q$ then by the above matrix the result is the false proposition:

$$(0 \; \varepsilon \; \text{Nn}) \cdot (\text{Mz} \; \varepsilon \; \text{G})$$

(i.e. "0 is a number and Mozart is a Greek")

On the other hand, by substituting (0 ε Nn) for P and ~(Mz ε G) for Q, we have as a result, the true proposition:

$$(0 \; \varepsilon \; \text{Nn}) \cdot {\sim}(\text{Mz} \; \varepsilon \; \text{G})$$

(i.e. "0 is a number and Mozart is not a Greek")

1.24. *"Or"*

We use the sign v (read "vel") *between* formulae, and it is defined by the following matrix:

P	Q	$P \lor Q$
T	T	T
F	T	T
T	F	T
F	F	F

The matrix shows that if vel occurs between two formulae each of which represents a false proposition, the result is a formula representing a false proposition. In every other possible case the result is shown to be a formula representing a true proposition. For example, if we substitute (0 ε Nn) for P and (Mz ε G) for Q in the formula P v Q, then by the above matrix, the result is the true proposition:

$$(0 \ \varepsilon \ \text{Nn}) \ \text{v} \ (\text{Mz} \ \varepsilon \ \text{G})$$

(i.e. "0 is a number or Mozart is a Greek")
Alternatively, substituting ~(0 ε Nn) for P and (Mz ε G) for Q we have as a result the false proposition:

$$\sim(0 \ \varepsilon \ \text{Nn}) \ \text{v} \ (\text{Mz} \ \varepsilon \ \text{G})$$

(i.e. "0 is not a number or Mozart is a Greek")

1.25. "If . . . then . . ."
We use the sign ⊃ (read "hook") *between* formulae and adopt it as an abbreviation for another expression- one involving constants whose matrices have already been given. The definition not being, in this case, a definition by matrix, the symbol "=df" is used (meaning "is a definitional abbreviation for"):

$$P \supset Q \ =df \ \sim(P \cdot \sim Q)$$

In other words, any proposition of the form "If . . . then - - -" is rendered in logic by a formula of the form $P \supset Q$, regardless of any relationship of meaning that may hold between the component propositions ". . ." and "- - -" in it. Clearly, this definition of the phrase is a very wide one. For, under it, the following propositions are true propositions:

$$(\text{Mz} \ \varepsilon \ \text{G}) \supset (0 \ \varepsilon \ \text{Nn})$$

(i.e. "If Mozart is a Greek then 0 is a number")

$$(\text{Mz} \ \varepsilon \ \text{G}) \supset \sim(0 \ \varepsilon \ \text{Nn})$$

(i.e. "If Mozart is a Greek then 0 is not a number")

$$(\text{Pl} \ \varepsilon \ \text{G}) \supset (0 \ \varepsilon \ \text{Nn})$$

(i.e. "If Plato is a Greek then 0 is a number")

Surprising as these results may seem, they do not detract from the usefulness of the definition in logic. These results are, in fact, surprising only when the definition is misunderstood as an attempt to express the full meaning of the phrase "if ... then - - -" as normally used. But any case of its normal use such as occurs in the proposition: "If he is a citizen then he is a voter" is covered by the definition, i.e.:

$$(x \ \varepsilon \ C) \supset (x \ \varepsilon \ V)$$

We can therefore safely give, as a negative defence of this definition, that the phrase "if ... then - - -" never occurs in ordinary speech without our defining conditions being satisfied; even if, in ordinary occurrences of the phrase, other conditions happen to be satisfied as well.

1.26. *Equivalence*

We use the sign $\equiv$ (read "three bars") *between* formulae and, as in the case of the hook sign, adopt it as a means of abbreviation. It is defined as follows:

$$P \equiv Q \ =\!df \ (P \supset Q) \cdot (Q \supset P)$$

By the definition of hook and the matrices for dot and for tilde, this definition entails that if three bars occurs between two formulae, each of which represents a true proposition, or between a pair representing false propositions, the result is a formula representing a true proposition. In the remaining two cases the result is a formula representing a false proposition. This can be seen from the following table:

P	Q		$P \equiv Q$		
		$\sim (P \cdot \sim\!Q)$	$\cdot$	$\sim (Q \cdot \sim\!P)$	
T	T	T F F	T	T F F	
F	T	T F F	F	F T T	
T	F	F T T	F	T F F	
F	F	T F T	T	T F T	
		1 c a	3	2 d b	

Column 3 is calculated as follows: column (a) consists of the values which we are directed (by the tilde matrix) to assign to ~Q, taking each of the possible values of Q in turn. Column (b) gives the result of proceeding in the same way for ~P. Column (c) consists of the values which we are directed (by the dot matrix) to assign to $(P \cdot$ ~$Q)$, taking each of the possible values of its components in turn (as given in column (a) and the column for P). Column (d) gives the analogous result for $(Q \cdot$ ~$P)$. Column (1) consists of the values which we are directed (by the tilde matrix) to assign to ~$(P \cdot$ ~$Q)$, taking each of the possible values in column (c) in turn. Column (2) gives the analogous result for ~$(Q \cdot$ ~$P)$. Column (3) consists of the values which we are directed (by the dot matrix) to assign to the entire formula, taking each of the possible pairs of values of its two main components in turn (as given in columns (1) and (2)). As our final result, column (3) shows that only when P and Q are either both true or both false, is P equivalent to Q according to our definition. In short the formula $P \equiv Q$ means "P is true if and only if Q is true".

Again, we have such superficially surprising results as that the following propositions are true propositions:

$$(\text{Pl } \varepsilon \text{ G}) \equiv (0 \ \varepsilon \text{ Nn})$$

(i.e. "Plato is a Greek, if and only if 0 is a number")

$$(\text{Mz } \varepsilon \text{ G}) \equiv \text{~}(0 \ \varepsilon \text{ Nn})$$

(i.e. "Mozart is a Greek, if and only if 0 is not a number") For reasons analogous to those given above, these results do not detract from the value of this definition in logic.

It should, however, be noted in this connection that if we have a proposition, say P, from which we are able to deduce another proposition, say Q, and vice versa, the two propositions are equivalent in a stronger sense than that defined above. This follows from the fact that if Q is deducible from P we have a case of "If . . . then - - -" where the phrase is being used in an important way, which is also stronger in meaning than the sign "$\supset$". For in such a case, we are able to claim not merely:

$$\text{~}(P \cdot \text{~}Q)$$

but that given that P is true, it is a matter of logic that Q is not false. Analogously, where P and Q are mutually deducible from one another, we are able to claim not merely:

$$\sim(P \cdot \sim Q) \cdot \sim(Q \cdot \sim P)$$

but that given that P is true, it is a matter of logic that Q is not false, *and vice versa*. Thus the sign "$\supset$" can be read as *logically entails* in case that we have been able to show how Q can be *deduced from* P; and the sign "$\equiv$" can be read as *logically entail one another*, in case that we have been able to show how Q can be deduced from P *and* how P can be deduced from Q.

Since the proof of a theorem of the form "If . . . then - - -" consists in showing how the second component proposition "- - -", can be deduced from the first ". . .", such a theorem, when proved, is a very much stronger claim than:

$$(\ldots) \supset (\text{- - -}) \quad \text{or} \quad \sim((\ldots) \cdot \sim(\text{- - -}))$$

It is, in fact, the claim that ". . ." logically entails "- - -". This, then, is the interpretation to be given to the sign "$\supset$" in the context of a proof.

1.27. "*All*"

We use the sign "(x)", (read "for all x") *immediately to the left of a propositional function*, to mean that the result of substituting any (no matter which suitable) value for the variable x in that propositional function, will be a true proposition. For example, if we use G as before and Ml for the class of mortals, we can construct the propositional function:

$$(x \; \varepsilon \; G) \supset (x \; \varepsilon \; Ml)$$

Now the range of values of the variable in this function is all human beings. For only with such values could we obtain a true or false (as opposed to a meaningless) proposition from this propositional function. Thus the following are in the range of values concerned:

Plato
Socrates
Mozart

If we substitute each of these values in turn for the variable x in the above propositional function, we have the following three propositions:

$$(\text{Pl } \varepsilon \text{ G}) \supset (\text{Pl } \varepsilon \text{ Ml})$$
$$(\text{St } \varepsilon \text{ G}) \supset (\text{St } \varepsilon \text{ Ml})$$
$$(\text{Mz } \varepsilon \text{ G}) \supset (\text{Mz } \varepsilon \text{ Ml})$$

From the well-known nationalities of these individuals of the past and the definition of the sign "$\supset$", it follows that each of these propositions is true. The claim that any such proposition is true, namely the proposition:

"All Greeks are mortal"

is formulated in logic by prefixing the sign "(x)" to the propositional function from which they were obtained, thus:

(IV) $(x)[(x \, \varepsilon \, \text{G}) \supset (x \, \varepsilon \, \text{Ml})]$

Notice that we bracket off the propositional function before prefixing the sign "(x)". This is necessary in order to distin guish between (IV) and, say,

$$(x)(x \, \varepsilon \, \text{G}) \supset (\text{Mz } \varepsilon \text{ Ml})$$

(i.e. "If Everything is a Greek then Mozart is a mortal")
Now (IV) will be true as long as the propositional function:

$$(x \, \varepsilon \, \text{G}) \cdot \sim(x \, \varepsilon \, \text{Ml})$$

does not give rise to any true proposition, for (IV) is the same proposition as:

(V) $(x)\sim[(x \, \varepsilon \, \text{G}) \cdot \sim(x \, \varepsilon \, \text{Ml})]$

by the definition of the sign "$\supset$". Thus, if there is *at least one* Greek who is *not* mortal (IV) and (V) will represent a false proposition. In that case the assertion that there is *not* at least one Greek who is *not* mortal, is the same assertion as that all Greeks *are* mortal. We make use of this fact in our definition of the word "some".

1.28. *"Some"*

We use the sign "$(\exists x)$" (read "for some x") *immediately to the left of a propositional function*, to mean that there exists *at least one* value in the range of appropriate values of the variable x in that propositional function, which gives rise to a true proposition when substituted for x. For example, if we use Pr to name the class of prime numbers we can construct the propositional function:

$$(x \ \varepsilon \ \mathrm{Nn}) \cdot (x \ \varepsilon \ \mathrm{Pr})$$

Here, the following individuals are in the range of values concerned:

$$1$$
$$2$$
$$3$$
$$4$$
$$\ldots \text{etc.}$$

If we substitute each of these in turn for the variable x in the above propositional function, we have the following propositions:

$$(1 \ \varepsilon \ \mathrm{Nn}) \cdot (1 \ \varepsilon \ \mathrm{Pr})$$
$$(2 \ \varepsilon \ \mathrm{Nn}) \cdot (2 \ \varepsilon \ \mathrm{Pr})$$
$$(3 \ \varepsilon \ \mathrm{Nn}) \cdot (3 \ \varepsilon \ \mathrm{Pr})$$
$$(4 \ \varepsilon \ \mathrm{Nn}) \cdot (4 \ \varepsilon \ \mathrm{Pr})$$

From the definition of "prime number" and the definition of "·", it follows that the first three of these propositions are true and the last one, false. The claim that at least one such proposition is true, namely the proposition:

"Some numbers are prime numbers"

is formulated in logic by prefixing the sign "$(\exists x)$" to the propositional function from which they were obtained, thus:

(VI) $(\exists x)[(x \ \varepsilon \ \mathrm{Nn}) \cdot (x \ \varepsilon \ \mathrm{Pr})]$

Notice, again, that we bracket off the propositional function before prefixing the sign "$(\exists x)$".

Now if we reformulate (VI) with G and Ml in place of Nn and Pr respectively, we get the proposition:

(VII) $(\exists x)[(x \ \varepsilon \ \mathrm{G}) \cdot (x \ \varepsilon \ \mathrm{Ml})]$

(i.e. "Some Greeks are mortal")

If we now substitute $\sim(x \; \varepsilon \; \mathrm{Ml})$ for $(x \; \varepsilon \; \mathrm{Ml})$ in (VII) we obtain:

(VIII) $(\exists x)[(x \; \varepsilon \; \mathrm{G}) \cdot \sim(x \; \varepsilon \; \mathrm{Ml})]$

 (i.e. "Some Greeks are not mortals")

Then, in virtue of the equivalence explained at the end of I.1.27 we have:

$$\sim(\exists x)[(x \; \varepsilon \; \mathrm{G}) \cdot \sim(x \; \varepsilon \; \mathrm{Ml})] \equiv (x)\sim[(x \; \varepsilon \; \mathrm{G}) \cdot \sim(x \; \varepsilon \; \mathrm{Ml})]$$

and, therefore, by double negation on the left-hand side (after negating both sides, as permitted in I.1.26):

$$(\exists x)[(x \; \varepsilon \; \mathrm{G}) \cdot \sim(x \; \varepsilon \; \mathrm{Ml})] \equiv \sim(x)\sim[(x \; \varepsilon \; \mathrm{G}) \cdot \sim(x \; \varepsilon \; \mathrm{Ml})]$$

For this reason we are able to define "some" in terms of "all", as follows:

$$(\exists x)\phi \; =df \; \sim(x)\sim\phi$$

where the symbol ϕ is being used to signify any propositional function in which the variable in question (here, x) occurs. We could, of course, equally well have defined "all" in terms of "some". (*Note:* the symbols "(x)" and "$(\exists x)$" are known as *quantifiers*, the former being the *universal quantifier*, and the latter the *existential quantifier*.)

1.3 The Use of Quantifiers in Proofs

The theorems with which we shall be mostly concerned are of two kinds, namely, those which assert the existence of some class, and those which assert a perfectly general proposition such as:

$$(x)[(x \; \varepsilon \; \mathrm{A}) \supset (x \; \varepsilon \; \mathrm{B})]$$

The proof procedure in this latter case is as follows: we assume the propositional function:

$$(x \; \varepsilon \; \mathrm{A})$$

as an hypothesis (*hyp.*); thus treating it as if it were a proposition. We then list the steps required to show that it entails the propositional function:

$$(x \; \varepsilon \; \mathrm{B})$$

treating this also as if it were a proposition. We then argue, in

effect, that if the first propositional function had been a true
proposition about any individual whatever (in the appropriate
range), the second propositional function would have been a
true proposition about that individual. From this we conclude
that the propositional function:

$$(x \ \varepsilon \ A) \supset (x \ \varepsilon \ B)$$

holds universally, that is:

$$(x)[(x \ \varepsilon \ A) \supset (x \ \varepsilon \ B)]$$

In the case of existence theorems (i.e. assertions of the form
"For some x . . .") the procedure is to list the steps required to
show that from given axioms and definitions, there follows
logically a propositional function (say: $x \ \varepsilon \ Pr$) which, had it
been a proposition about a specified individual (say: the
number 3), would have entitled us to the required existence-
claim. From this we conclude that the propositional function
in question holds in at least one case, that is:

$$(\exists x)(x \ \varepsilon \ Pr)$$

Furthermore, from the existence-claim expressed by a
propositional function having an existential quantifier prefixed,
it is perfectly legitimate to proceed to an analogous but un-
quantified assertion using an arbitrarily chosen variable. For
example, from:

$$(\exists x)(x \ \varepsilon \ A)$$

we may, in any proof, infer say:

$$z \ \varepsilon \ A$$

as long as the variable z has not been used earlier in the proof.
(If it had the choice of it would not, of course, have been
entirely arbitrary.) In this way we are able, as in proofs we
need to be able, to make assertions about a supposed individual,
even when we are not in a position to specify the actual individual in question.

Analogously, we can always infer from the premise that a
propositional function holds universally, the conclusion that it

holds for any particular individual with which we may be concerned. For example, from:

$$(x)(x \ \varepsilon \ A)$$

we may infer:

$$k \ \varepsilon \ A$$

where k is the name of a particular individual. Or we may, if we wish, merely infer a propositional function, say,

$$z \ \varepsilon \ A$$

where z is any variable we like to choose, whether or not we have made use of it earlier in the proof.

This concludes our preliminary survey of the symbolism of logic used in the pages which follow.

OPERATIONS ON CLASSES

2.1. Classes and Composite Wholes

THE question, "What is a class?", is not an easy one to answer, nor is it the purpose of set theory to find an answer to it. For in set theory the term *class* is left undefined. But in set theory, as in any other formal deductive system, the undefined terms have an intended interpretation. Some idea of what this intended interpretation is, then, is essential to an understanding of what set theory is all about. We have, therefore, to explain what a class is, without explicitly defining it.

The easiest way to do this is to indicate by means of examples the peculiar force of the word "class". In the first place, the word "class" is correlative with the phrase "member of", just as the word "thing" (or "whole", or "whole thing") is correlative with the phrase "part of". Now the peculiar force of the word "class" can be seen from the fact that these two pairs of correlatives are not alternative locutions. For instance, from the fact that a stone is a part of a wall and the fact that the wall in question is part of a castle, we may infer that the stone is part of the castle. Yet we may not infer from the fact that Mr X is a member of the British Nation and the fact that the British Nation is a member of the United Nations, that Mr X is a member of the United Nations. Thus, in general, a part of a part of a whole is *always* said to be a part of that whole, whereas a member of a member of a class is *not* always (although in certain cases it may be) said to be a member of that class. To this extent at least, then, the words "class" and "whole" have distinct uses and are therefore not synonymous.

Whatever else a class is, then, it is not simply a whole of parts. We can, however, be more positive about the nature of a class, than this. For there is some measure of common agreement as to when two references to classes are to be taken as references to the same class and when not. For example, the

16

class made up of all the books read by one student during a given year is unlikely to be the same as that made up of all the books read by some other student during that year. If one is a science student and the other an arts student, the two classes will probably be entirely distinct. If both are reading the same or related subjects, on the other hand, the classes are likely to overlap; and it is just conceivable, if both are attending the same course, that the two classes may be identical. But in each case the identity or otherwise of the two classes is determined by whether or not the conditions-for-membership of the one class is fulfilled by at least one book which fails to fulfil the conditions-for-membership of the other class. In general, two class-references are references to different classes only so long as there is something to which the predicate of the one reference applies, and to which the predicate of the other reference does not. We may, therefore, take a class to be simply the field of application (or extension) of a predicate. We shall do so, with the added proviso that this is not an explicit definition and that it is subject to further qualification (cf. II.0.2, p. 38).

2.2. Conventions for Diagramming Class-structure

One reason why many people find set theory difficult is that they cannot form any kind of mental picture of what lies behind the intricate manipulations of the symbols. This can be overcome to a large extent with the help of diagrams in the early stages. Unfortunately, the diagrams in most common use for this purpose (e.g. the Venn diagrams) do not provide for a clear and unambiguous visible means of distinction between class-inclusion and class-membership. A thorough grasp of this distinction is essential to an understanding of everything that follows. It can be illustrated by considering the inference already mentioned as invalid, from (i) "He is a member of the British Nation" and (ii) "The British Nation is a member of the United Nations", to: "He is a member of the United Nations". For we may contrast with this the inference from, for instance, (i) "The class of Greeks is included in the class of men", and (ii) "The class of men is included in the class of mortals", to: "The class of Greeks is included in the class of mortals". This is clearly a valid inference. Thus, in general, a sub-class of a sub-class of a given class is a sub-class of (or included in) the given class; whereas a member of a member of a given class is not, in general, a member of the given class. Now the most common method of indicating by a diagram the case in which one class y is included in another x, is the use of two circles, either concentric or overlapping, thus:

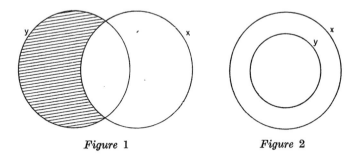

Figure 1   Figure 2

In Fig. 1 the shading indicates absence of members in the class indicated by the shaded region. But we might wish to indicate diagrammatically that a certain class (say, C) has as *members* two *classes* (say x and y) one of which is *included in* the other. Using the same method, we should have to choose between the two following diagrams:

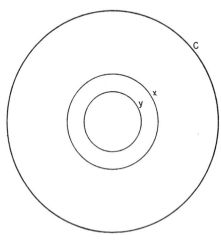

Figure 3

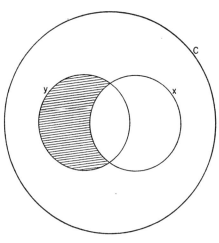

Figure 4

But in neither case is it clear simply from inspecting the diagram which of the classes are related by *inclusion* and which by *membership*.

We shall therefore adopt the following conventions:

(i) A class shall be represented thus:

○─────────────────── ─────────

(ii) Where C is a class, the fact that C has, say, x,y,z, as members, and no other members, shall be represented thus:

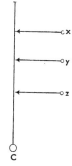

Figure 5

(*Note:* all the classes we shall be concerned with will be, like *C* above, classes of classes.)

(iii) Where *C* and *D* are classes, the fact that *C* has, say, *x,y,z* as members and that *C* is itself a member of *D*, shall be represented thus:

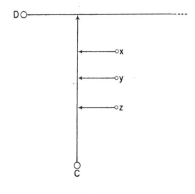

Figure 6

(*Note:* The three dots serve to show that the class in question may have members other than the one or more indicated.)

(iv) Where *C* is a class, the fact that *C* is a member of, say, *D* and *E*, may be represented either by:

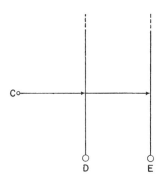

Figure 7

or by:

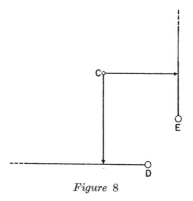

Figure 8

Using these conventions we can indicate the inclusion of one class in another by the following diagram:

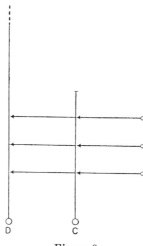

Figure 9

Here, the class C is included in the class D (or: $C \subseteq D$). In this diagram, then, we have some indication at the intuitive level, of what lies behind the standard definition of class-inclusion, namely:

$$C \subseteq D \ =df \ (x)(x \ \varepsilon \ C \supset x \ \varepsilon \ D)$$

In what follows we shall append to each diagram the formula whose intended interpretation is thus illustrated.

2.3. Elementary Operations

By an **operation on a class** or classes, we mean the definition of a new class in terms of a given class or classes. For example, given two classes, say C and D, we can formulate the predicate: "... is a member of both C and D" thereby enabling reference to be made to a new class, whose membership is determined in a particular way by the membership of C and D.

We now proceed to name, illustrate and define some of the more important operations on classes with which we shall be concerned.

Complement of a Class

The complement of a class, say C, is the class defined by: "... is not a member of C". Thus we have:

$$\overline{C} =df \text{ the class defined by} : [\sim(x \; \varepsilon \; C)]$$

Proper Sub-class

From the above definition of inclusion it can be seen that "C is included in D" does not rule out the possibility that also "D is included in C". Indeed, where both are the case we have the defining condition for the identity of C and D. Where C is included in D but D is not included in C, we say that C is a proper sub-class of D. This may be represented as follows:

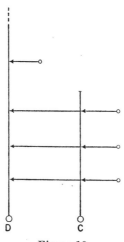

Figure 10

$$C \subset D \ =df \ [C \subseteq D \cdot \sim(D \subseteq C)]$$

(*Note:* This is not strictly speaking an operation since $C \subset D$ is a statement, not the name of a new class.)

Product

The product (or intersection) of two classes C and D is the class defined by: ". . . is a member of C and is a member of D". This may be represented as follows:

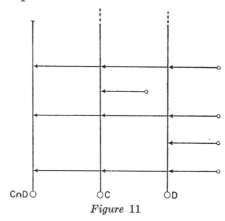

Figure 11

$$C \cap D \ =df \ \text{the class defined by: } [x \ \varepsilon \ C \cdot x \ \varepsilon \ D]$$

Sum

The sum of two class C and D is the class defined by: "... is a member of C or of D". This may be represented as follows:

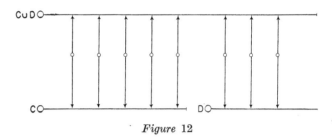

Figure 12

$$C \cup D \ =df \ \text{the class defined by:} \ [x \ \varepsilon \ C \ \mathrm{v} \ x \ \varepsilon \ D]$$

Manifold Sum

The sum of the elements (or manifold sum) of a class C of classes is the class of members of members of that class. It is, therefore, the class defined by: "... is a member of some member of C". This may be represented as follows:

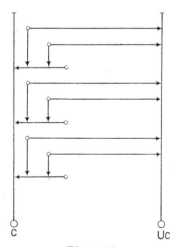

Figure 13

$$\mathsf{U}C \ =df \ \text{the class defined by:} \ [(\exists z)(z \ \varepsilon \ C \cdot x \ \varepsilon \ z)]$$

Manifold Product

The manifold product (or intersection of the elements) of a class C of classes is the class of members of every member of that class. It is, therefore, the class defined by: "... is a member of every member of C". This may be represented as follows:

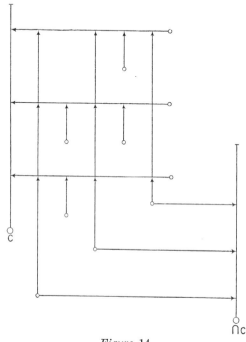

Figure 14

$\cap C$ =df the class defined by: $[(z)(z \; \varepsilon \; C \supset x \; \varepsilon \; z)]$

Unit-class and Pair-class

The unit class of a class C is the class whose sole member is C; similarly the pair-class of the classes C and D is the class whose sole members are C and D. The unit-class of C is, therefore, defined by: "... is identical with C", and the pair-class of C and D, by: "... is identical with C or is identical with D". These may be represented as in Fig. 15.

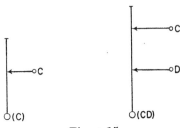

Figure 15

(C) $=df$ the class defined by: $[x = C]$

(CD) $=df$ the class defined by: $[x = C \lor x = D]$

Ordered Pair

An ordered pair is a pair of classes in a given order and such that it differs from a pair having the same members in a different order. We shall expand this statement more fully later (cf. II.0.11) confining ourselves here to an intuitive grasp of the concept. For this purpose we add to our conventions for diagramming class-structure, the following rule:

(v) The fact that a given class, say B, has C and D as its sole members *and that* membership in B is dependent on the order in which C and D are placed, shall be represented thus:

$$B \!-\!\!\!\!-\!\!\!\!-\!\!\!\!-\!\!\!\!-\!\!\!\!-\!\!\!\!-\!\!\!\!-$$

(*Note:* We use the notation $\{CD\}$ to distinguish an ordered pair from a pair.)

$\mathrm{mem}_1 A$

The class $\mathrm{mem}_1 A$ is a class whose members are those ordered pairs whose first member is a member of A. It is therefore the class defined by: "... is an ordered pair whose first member is a member of A". This may be represented as follows:

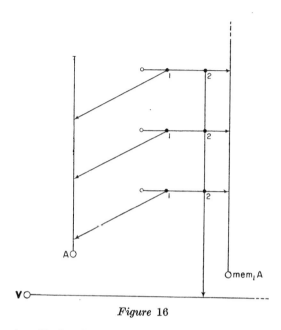

Figure 16

mem_1A $=df$ the class defined by: $[(\exists zy)(\{zy\} = x \cdot z \,\varepsilon\, A)]$

(*Note:* V, here, represents the entire universe of individuals, to *every* one of which *each* member of A is paired.)

Domain

The domain of a class C of ordered pairs is the class whose members are first members of members of C. It is, therefore, the class defined by : ". . . is a first member of an ordered pair which is a member of C". This may be represented as follows:

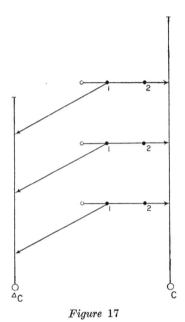

Figure 17

$$\triangle C \ =df \ \text{the class defined by} : [(\exists y)(\{xy\} \ \varepsilon \ C)]$$

Converse

The converse of a class C of ordered pairs is the class whose members are those ordered pairs such that if the order of their members were reversed they would be members of C. It is, therefore, the class defined by: ". . . is an ordered pair $\{xy\}$ such that $\{yx\}$ is a member of C". This may be represented as in Fig. 18:

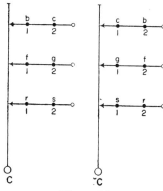

Figure 18

$\smallsmile\!C$ =df the class defined by: $[(\exists yz)(\{yz\} = x \cdot \{zy\} \, \varepsilon \, C)]$

Converse Domain

The converse domain of a class C of ordered pairs is the class whose members are first members of the converse of C. Thus the converse domain of C is the domain of the converse of C. It is therefore the class defined by: "... is a first member of an ordered pair which is a member of the converse of C". This may be represented as follows:

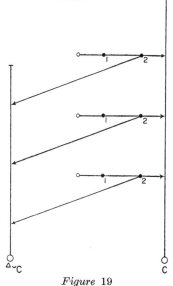

Figure 19

$\vartriangle\!\smallsmile\!C$ =df the class defined by: $[(\exists y)(\{xy\} \, \varepsilon \, \smallsmile\!C)]$

The Null Class

Since a predicate may be such that it applies to nothing, we may speak of a class having no members. Since all such classes have the same members (namely none) we may speak of *the* empty (or null) class. It is the class defined by: "... is not identical with itself" and is represented by the symbol: $\wedge$. Thus we use the notation, $\wedge$, for any class whose defining condition is such that nothing satisfies it.

Self-augment of a Class

We shall use Quine's term "self-augment" (see: Quine, *Mathematical Logic*, p. 247), to mean the class whose members are either a given class itself or the members of that class. The self-augment of a class C is, therefore, the class defined by: "... is identical with C or is a member of C". This may be represented as follows:

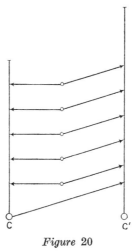

Figure 20

$$C' =_{df} \text{ the class defined by: } [x = C \vee x \, \varepsilon \, C]$$

$\frac{A}{B}\mathrm{Pr}$

The class $\frac{A}{B}\mathrm{Pr}$ is the class whose members are those ordered pairs whose first members are members of A and whose second members are members of B. It may be read "$A-B$ pair", and is defined by: "... is an ordered pair which is a member of the product of mem_1A and mem_2B". (Here, mem_2B is to be understood by analogy with the explanation of mem_1A). This may be represented as follows:

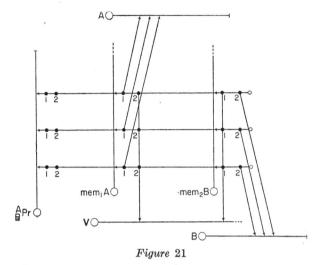

Figure 21

$\frac{A}{B}Pr$ $=df$ the class defined by: $[mem_1A \cap mem_2B]$

This completes our preliminary survey of the more important of the operations which we shall be using. Familiarity with them is essential to an understanding of Part II.

THE BERNAYS THEORY OF
FINITE CLASSES AND FINITE SETS

FOREWORD TO PART TWO

THE purpose of Part Two is to show how the symbolism explained in Part I can be used to set up, in a relatively formal manner, a deductive system of truths about classes, from which the elementary truths of mathematics can be derived. This could be done in a number of alternative ways. For there are many different systems of this kind from which to choose. The system given below is adapted from part of Paul Bernays' *A System of Axiomatic Set Theory*[1] which is admirably suited to the present purpose. For the ideas and reasoning involved are, basically, very simple. As Bernays himself says: "it adopts the principal idea of von Neumann, that the elimination of the undefined notion of a *property* . . . which occurs in the original axiom system of Zermelo, can be accomplished in such a way as to make the resulting axiom system elementary . . ." (*op. cit.*, Pt. I, p. 65).

In the following pages a small part of the Bernays System has been expanded and set into a formal framework. This framework has been devised with the sole aim of combining, as far as possible, simplicity with formality. The result neither is nor is intended to be a completely rigorous formalization of the Bernays System. The definitions and the proofs of the major theorems are, for the most part, symbolizations of the definitions and proofs given by Bernays. Many of the other proofs, however, may well fall short of the elegance with which Bernays himself would have formally proved them. But this presentation will, it is hoped, serve as a first step to an understanding of the more rigorously formalized systems of set theory.

[1] See: References, p. 103.

INTRODUCTION

0.1. Pairs, Ordered Pairs and k-Tuplets

IN the Bernays system a distinction is drawn between two senses of the word "class". The distinction is expressed by using the word "class" itself in one particular sense and by using, in addition, the term "set". The point of the distinction will be explained later (see II.0.2, p. 38). Now the theory of pairs, ordered pairs and k-tuplets which is a fundamental part of the system, concerns operations which are confined to sets. The nature of these operations is, however, quite independent of the domain of discourse to which they are applied. We shall, therefore, use the word "set" *ab initio*, despite the fact that the distinction between a set and a class has yet to be explained.

0.11. *Pairs and Ordered Pairs*

It will be shown in the next section that sets whose only member is some given set, and sets whose only members are either some given set or another given set, exist for this system. Such sets will be denoted in the usual way by (x) and (xy), respectively, where x,y, are variables for sets. (See above, pp. 25–26.) It will also be shown that coextensionality (i.e. having the same membership) determines identity between sets as well as between classes. Thus **ordered pairs,** in the Kuratowski[1] sense are definable in the system, i.e.:

$$\{ab\} \; =df \; ((a),(ab))$$

As an immediate consequence, we have:

$$\{ab\} = \{cd\} \; \equiv \; a = c \cdot b = d$$

For, either $c = d$ or $c \neq d$. But, if $c = d$, $\{cd\} = ((c),(cd))$ $= ((c)(c))$, and $((a),(ab))$ is coextensive with $((c))$. So that $(a) = (c)$, $(ab) = (c)$ and $(a) = (ab)$. Therefore $b = a$ and:

---

[1] *Fundamenta Mathematica*, 2; 1921. See also: QUINE, *Mathematical Logic*, section 36.

$a = c$ and $b = d$. Similarly, if $c \neq d$ then $a \neq b$ and: $a = c$ and $b = d$.

(*Note:* We shall refer to ordered pairs simply as "pairs" unless the context demands otherwise.)

0.12. k-Tuplets

With ordered pairs available in the system, the concept of a **k-tuplet** is introduced in terms of *ordered pair* as, a set formed out of the sets $a_1, \ldots, a_k$ (to be called its members) in the following way:

(1) A set b is itself a 1-tuplet formed out of b.

(2) A $(k+1)$-tuplet formed out of the sets $b_1, \ldots, b_{k+1}$ is a pair $\{rs\}$ such that r is a p-tuplet, s is a q-tuplet, and p + q is the same as k + 1.

Thus a 2-tuplet formed out of b and c is simply the ordered pair $\{bc\}$. It should be noticed that any $(k+1)$-tuplet is also a k-tuplet, but that its members as a k-tuplet differ from its members as a $(k+1)$-tuplet.

Three further related concepts are required in the system. These are:

(1) Schema of a k-tuplet.

(2) Normal k-tuplet.

(3) Correspondence between a k-tuplet and a system of values of given variables.

The definitions of these are as follows:

(1) The *schema of a k-tuplet* is the result of replacing its members in order by variables $b_1, \ldots, b_k$. Thus the schema of the k-tuplet:

$$\{\{\{cd\}g\}h\}$$

is:

$$\{\{\{b_1 b_2\} b_3\} b_4\}$$

(2) We define a k-tuplet as normal in terms of the degree of each of its members, and the latter is defined in terms of the schema of a k-tuplet as follows:

The *degree of a variable* in the schema of a k-tuplet is the number of pairs of brackets enclosing it; the degree of a member of a k-tuplet is the degree of the corresponding variable in the schema.

A *normal k-tuplet* is a k-tuplet in which every member b_i (i less than k) in $b_1, \ldots, b_k$ has degree i, and b_k has degree k-1. Thus the k-tuplet given above is not normal, but would have been had it been bracketed as follows:

$$\{c\{d\{gh\}\}\}$$

We shall use the expression: $^kT: z$, to mean: "z is a normal k-tuplet" and define it as follows:

$$^kT: z =_{df} [(\exists x_1, \ldots, x_k)(z = \{x_1\{x_2, \ldots \{x_{k-1}x_k\}, \ldots\}\}) \cdot (x)(\{x\} = x)]$$

(3) The k-tuplet *corresponding to a system of values* $b_1, \ldots, b_k$ is the k-tuplet whose members are $b_1, \ldots, b_k$ in that order.

k-tuplets may differ either in the position of their brackets alone, as in the case of the above two examples, or in the order of their members alone, or in both ways. The following examples illustrate the latter two alternatives:

(i)	$\{b\{\{dc\}a\}\}$
(ii)	$\{b\{a\{dc\}\}\}$
(iii)	$\{\{ba\}\{dc\}\}$
(iv)	$\{\{ab\}\{cd\}\}$
(v)	$\{a\{b\{cd\}\}\}$

Here, (iii) and (iv) differ in the order of their members alone, whereas (i) and (ii) for instance, differ both in the order of their members and in the position of their brackets. It should also be noticed that the alterations made between (i) and (v) have the effect of transforming a quadruplet into a normal quadruplet. But it has, of course, yet to be shown that for any given k-tuplet there exists the corresponding normal k-tuplet.

0.2. Class Conditions

Although the term *class* is being used to mean the extension of a predicate, we do not assume that every predicate determines a class. For, if we assumed this then, as Russell has shown (see: *Introduction to Mathematical Philosophy*, p. 136) we have a paradox on our hands. The paradox would arise in the following way. The predicate "is a class" would determine a class, say U, which itself satisfied this predicate, and U would belong to U. Thus we should also have the predicate "is a class which belongs to itself" determining a class to which U belonged. We would then have to allow that the predicate "is a class which does not belong to itself" determined a class, say K. In that case, if K belonged to K, it would thereby satisfy this predicate and thus not belong to itself. Yet if K did not belong to itself, then it would satisfy this predicate and thus belong to K. In short, K would belong to itself, if and only if it did not belong to itself. The way in which this paradox is avoided in the present system is, in outline, as follows. We first of all reject the assumption that every predicate determines a class. But since there are many classes for which we can specify a determining predicate, a distinction is introduced between a class—the use of this term presupposing that a determining predicate is available, and a set—meaning by this, the only type of entity which a class can have as a member. This distinction is built into the system by means of a device for ensuring that no predicates other than predicates of sets (i.e. assertions whose only variables are variables for sets) shall, in our system, operate as determining conditions for classes. Thus the only classes with which the system can deal are classes of sets (as opposed to classes of classes). Clearly, the unwanted class K cannot now be constructed. Our aim, therefore, will be to show that, despite a restriction to classes of sets, we can construct all the classes we require, while at the same time entirely excluding the unwanted class (and certain others like it) from our range of discourse.

0.21. *Authorized Predicates*

The following notation will be used:

$k \, \beta \, A$ for: the relationship of *being an element of* between a *set* k and a *class* A, *in that order*.

$[\phi]^{v_1, \, \ldots, \, v_k}_{\gamma_1, \, \ldots, \, \gamma_n}$ for: the well-formed formula ϕ with some or all of $v_1, \ldots, v_k$ among its free set variables and $\gamma_1, \ldots, \gamma_n$ as parameters.

$\vdash^\wedge[\phi]^{v_1, \, \cdots, \, v_k}_{\gamma_1, \, \cdots, \, \gamma_n}$ for:

$$(\gamma_1, \, \cdots, \, \gamma_n)(\exists A)(v_1, \, \ldots, \, v_k)[(\phi \equiv \{v_1\{v_2 \ldots \{v_{k-1}v_k\} \ldots\}\} \, \beta \, A)$$
$$\cdot \, (z)(z \, \beta \, A \supset {}^kT{:}\, z)]$$

By the last of the above definitions, the notation $\vdash^\wedge$ signifies that the expression in the square brackets following it, is a class condition on the set variables indicated, in the sense given by the definiens. That is, the formula ϕ in which $v_1, \, \ldots, \, v_k$ are free variables for sets and $\gamma_1, \, \ldots, \, \gamma_n$ are the parameters, determines a class each member of which is a normal k-tuplet corresponding to some system of values of the variables $v_1, \, \ldots, \, v_k$. Thus, where $k = 1$, we have:

$$\vdash^\wedge[\phi]^z \text{ for } (\exists A)(z)[(\phi \equiv z \, \beta \, A) \cdot (x)(x \, \beta \, A \supset {}^1T{:}\, x)]$$

For example, the assertion:

$$\vdash^\wedge[x = a]^x_a$$

means:

$$(\exists B)(x)[(x = a \equiv x \, \beta \, B) \cdot (z)(z \, \beta \, B \supset {}^1T{:}\, z)]$$

Or, where $k = 2$, for instance, the assertion:

$$\vdash^\wedge[x \, \beta \, A \cdot y \, \beta \, B]^{x,y}_{A,B}$$

means:

$$(\exists D)(x)(y)[(x \, \beta \, A \cdot y \, \beta \, B \equiv \{xy\} \, \beta \, D) \cdot (z)(z \, \beta \, D \supset {}^2T{:}\, z)]$$

In each such case, since coextensionality determines identity between classes, every authorized predicate determines a unique class.

0.22. *Rule for Defining Individual Constants*

So far we have a means of indicating that a given predicate is a class condition, but no means of permanent identification of the unique class determined by that predicate. In order to allow for this, we introduce the following Rule For Defining Individual Constants[1].

---

[1] cf. Suppes, *Introduction to Logic*, pp. 159–60.

If $(\exists!\, x)\phi$ *then* $(c = x) \equiv \phi$, *where* ϕ *has no free variables other than* x, *apart from parameters occurring in* c.

> *Note:* In the above rule, c is an expression for a class or a set; if ϕ has no parameters then c is a new individual symbol; if ϕ has one or more parameters then c consists of a new function symbol with those parameters as its (only) variables. Furthermore, the quantifier $(\exists!\, x)$ is defined as follows: $(\exists!\, x)\phi$ for: $(\exists x)(\phi \cdot (z)(\phi^* \supset z = x))$, where ϕ^* is like ϕ except that z occurs wherever x occurs in ϕ; that is, "There exists a *unique* x such that ϕ".

For example, given:

$$\vdash^{\frown}[x\,\beta\,A \cdot x\,\beta\,B]_{A,B}^{x}$$

then, introducing say, D (in accordance with II.0.21), and omitting the quantifier (in accordance with I.1.29) we can infer, as before:

$$(x)(x\,\beta\,A \cdot x\,\beta\,B \equiv x\,\beta\,\mathrm{D})$$

and since (introducing instead say G):

$$(x)(x\,\beta\,A \cdot x\,\beta\,B \equiv x\,\beta\,\mathrm{G})$$

entails:

$$(x)(x\,\beta\,\mathrm{D} \equiv x\,\beta\,\mathrm{G})$$

we have $\mathrm{G} = \mathrm{D}$ and we can, having thus proved the uniqueness of D, introduce, in accordance with our new rule, a particular individual constant, say $A \cap B$, by means of the equivalence:

$$A \cap B = \mathrm{D} \ \equiv\ (x)(x\,\beta\,A \cdot x\,\beta\,B \equiv x\,\beta\,\mathrm{D})$$

Again, given:

$$\vdash^{\frown}[x\,\beta\,A \cdot y\,\beta\,B]_{A,B}^{x,y}$$

we can infer:

$$(\exists \mathrm{D})(x)(y)[(x\,\beta\,A \cdot y\,\beta\,B \equiv \{xy\}\,\beta\,\mathrm{D}) \cdot (z)(z\,\beta\,\mathrm{D} \supset {}^{2}\mathrm{T}\colon z)]$$

from which it follows that:

$$(z)(z\,\beta\,\mathrm{D} \equiv (\exists x)(\exists y)(z = \{xy\} \cdot x\,\beta\,A \cdot y\,\beta\,B))$$

and since (introducing, say G):

$$(z)(z\,\beta\,\mathrm{G} \equiv (\exists x)(\exists y)(z = \{xy\} \cdot x\,\beta\,A \cdot y\,\beta\,B))$$

entails $G = D$, we can, having again proved the uniqueness of D, introduce a particular individual constant, say $\frac{A}{B}\mathrm{Pr}$, by means of the equivalence:

$$\frac{A}{B}\mathrm{Pr} = D \;\equiv\; (z)(z\,\beta\,D \equiv (\exists x)(\exists y)(z = \{xy\} \cdot x\,\beta\,A \cdot y\,\beta\,B))$$

Where, as in these examples, parameters occur, the new symbol is, of course, an individual constant (rather than a new function symbol) only if the parameters have their values fixed by reference to previous lines of the proof in which the new symbol is introduced.

In this way we are able, in proofs, to proceed in either direction between an assertion that some authorized predicate or class condition holds, and an assertion of membership in an identified class. For example, from:

(1) $(z\,\beta\,A \cdot z\,\beta\,B) \cdot \vdash^\frown [x\,\beta\,A \cdot x\,\beta\,B]^x_{A,B}$

we can proceed as follows:

(2) $(x)(x\,\beta\,D \equiv x\,\beta\,A \cdot x\,\beta\,B)$

(3) $z\,\beta\,D$ (1)

(4) $D = A \cap B$ (2)

(5) $z\,\beta\,A \cap B$ (3),(4).

(and vice versa).

For convenience, however, we shall, in the exposition of the system, abbreviate such moves by the use of the expression: "Let the class so defined be . . .", or by the annotation RIC ("Rule for Individual Constants").

THE BASIS OF THE SYSTEM

1.1. Primitive Constants

To PROCEED in a strictly formal way we should take as the basis of the system, the first-order function calculus with identity—this being the name given to a formalization of the logic informally explained in Part I. This, however, would be an unnecessarily complex approach for present purposes. Instead, we shall, in the steps of our proofs, make use only of the simple techniques already outlined (see especially I.1.29), taking these as the basic logic upon which the rest of the system is to be erected.

The first step in the actual construction of the system consists in the explicit formulation of the most fundamental among the axioms and definitions used in it. (Other axioms and definitions will be introduced at each successive stage of development, as they are required.) No definitions, however, can be introduced into a deductive system unless we already have some undefined terms (or "primitive constants") with which to express them. We begin, therefore, by exhibiting two symbols as primitive constants. These two constants are of crucial importance since the meaning of the word "set" is entirely determined by the relationships asserted by means of them, in the axioms of the system. They are called "special primitives" in the sense that they are additional to the primitive logical constants (tilde, dot, hook, etc.) occurring in the basic logic.

Special Primitives: ε β

The first of these, ε, is the familiar dyadic predicate (or two-term relation) of membership, used here, however, to designate the relation of *being an element of* between *sets* only. The second, β, is the dyadic predicate introduced in the previous section and intended as designating the relationship of *being an element of* between a *set* and a *class, in that order*.

For convenience we also adopt the following notational conventions to highlight the distinction between classes and sets:

(i) Variables for sets will be confined to lower-case letters.

(ii) Variables for classes will be confined to capitals.

In conjunction with these two conventions, we shall also adopt the following:

(iii) Free variables will be confined to: $a, \ldots, t, A, \ldots, T$.

(iv) Quantified variables will be confined to: $u, \ldots, z,$ $U, \ldots, Z$.

1.2. Initial Definitions

In the first of the following groups of definitions we show, on the left of the defined terms, an abbreviated form of expression which can be adopted for convenience, on the strength of conventions (i) and (ii) above.

(i)

$a \subseteq b$ for: $a \subseteq_{ss} b =df (z)(z \,\varepsilon\, a \supset z \,\varepsilon\, b)$ ("sub-set of a set")

$a \subseteq B$ for: $a \subseteq_{sc} B =df (z)(z \,\varepsilon\, a \supset z \,\beta\, B)$ ("sub-set of a class")

$A \subseteq b$ for: $A \subseteq_{cs} b =df (z)(z \,\beta\, A \supset z \,\varepsilon\, b)$ ("sub-class of a set")

$A \subseteq B$ for: $A \subseteq_{cc} B =df (z)(z \,\beta\, A \supset z \,\beta\, B)$ ("sub-class of a class")

$a \subset b$ for: $a \subset_{ss} b =df (a \subseteq_{ss} b \cdot \sim(b \subseteq_{ss} a))$ ("proper sub-set of a set")

$a \subset B$ (analogously)
etc.

(ii)

$a \leftrightarrow B$ $=df$ $(a \subseteq B \cdot B \subseteq a)$ ("a represents B")

rf: c $=df$ $(c \,\varepsilon\, c)$ ("c is reflexive")

tr: c $=df$ $(x,y)(x \,\varepsilon\, y \cdot y \,\varepsilon\, c \supset x \,\varepsilon\, c)$ ("c is transitive")

vct: c $=df$ $\sim(\exists x)(x \,\varepsilon\, c)$ ("C is empty")

$\sim$vct: c $=df$ $(\exists x)(x \,\varepsilon\, c)$ ("C is non-empty")

1.3. The Axioms of the System

1.31. *Formal Axioms* (FAx)

$a = b \supset (a \; \varepsilon \; c \supset b \; \varepsilon \; c)$

$a = b \supset (c \; \varepsilon \; a \supset c \; \varepsilon \; b)$

$a = b \supset (a \; \beta \; C \supset b \; \beta \; C)$

1.32. *Axioms of extensionality*

AxI

  (a) $(a \subseteq b \cdot b \subseteq a) \supset a = b$

  (b) $(A \subseteq B \cdot B \subseteq A) \supset A = B$

1.33. *Axioms of Direct Construction of Sets*

AxII

  (a) $(\exists x)(\text{vct}: x)$

    (There exists an empty set.)

  (b) $(z)(y)(\sim(z \; \varepsilon \; y) \supset (\exists s)(x)(x \; \varepsilon \; s \equiv x \; \varepsilon \; y \lor x = z))$

    (If y is a set and z is not in y, then there exists a set s which has y as a proper subset, and no members other than z and the members of y.)

1.34. *Axioms for Construction of Classes*

AxIII

  (a1) $\vdash\hat{\ }[c = a]_a^c$

  (a2) $\vdash\hat{\ }[\sim(c \; \beta \; A)]_A^c$

  (a3) $\vdash\hat{\ }[c \; \beta \; A \cdot c \; \beta \; B]_{A,B}^c$

  (b1) $\vdash\hat{\ }[(\exists y)(c = (y))]^c$

  (b2) $\vdash\hat{\ }[a \; \varepsilon \; b]^{a,b}$

  (b3) $\vdash\hat{\ }[a \; \beta \; A]_A^{a,b}$

  (c1) If $(z)(z \; \beta \; A \supset (\exists x,y)(z = \{xy\}))$ then $\vdash\hat{\ }[(\exists y)(\{cy\} \; \beta \; A)]_A^c$

  (c2) If $(z)(z \; \beta \; A \supset (\exists x,y)(z = \{xy\}))$ then $\vdash\hat{\ }[\{ba\} \; \beta \; A]_A^{a,b}$

  (c3) If $(z)(z \; \beta \; A \supset (\exists x,y,u)(z = \{x\{yu\}\}))$ then
      $\vdash\hat{\ }[(\exists x,y)(d = \{xy\} \cdot \{x\{yc\}\} \; \beta \; A)]_A^{d,c}$

1.35. *The Restrictive Axiom*

AxIV

$$(\exists x)(x \beta A) \supset (\exists x)(x \beta A \cdot \sim(\exists y)(y \beta A \cdot y \varepsilon x))$$

(If A is a non-empty class then it has at least one element which has no element in common with A.)

1.4. The Classes Admitted by AxIII

(a1) Let the class so defined (the unit-class of a set) be: $[a]$ (cf. Fig. 15, p. 26).

(a2) Let the class so defined (the complementary class of a class) be: $\bar{A}$ or: $\overline{/A/}$.

(a3) Let the class so defined (the intersection of two classes) be: $/A \cap B/$ (cf. Fig. 11, p. 23).

(b1) Let the class so defined (the class of unit sets) be: USt.

(b2) Let the class so defined (the class of membership pairs) be: MPr.

(b3) Let the class so defined (the class of first-member-in-A pairs) be: $\text{mem}_1 A$ or $\text{mem}_1/A/$ (cf. Fig. 16, p. 27).

(c1) Let the class so defined (the domain of a class of pairs) be: $\overset{\triangle}{A}$ or: $\overset{\triangle}{/A/}$ (cf. Fig. 17, p. 28).

(c2) Let the class so defined (the converse class of a class of pairs) be: $\overset{\smile}{A}$ or: $\overset{\smile}{/A/}$ (cf. Fig. 18, p. 29).

(c3) Let the class so defined (the coupling-to-the-left of a class of pairs) be: $\vec{A}$ or: $\overset{\rightarrow}{/A/}$.

Note: From the classes admitted by AxIII we may construct other classes as required. For example, we can construct the class "mem_1 USt" out of the classes $\text{mem}_1 A$ and USt, as follows:

(1) $(x)(x \beta \text{ mem}_1 A \equiv (\exists y,z)(x = \{yz\}$
$\cdot y \beta A))$ (AxIII(b3), p. 45)

(2) $(x)(x \, \beta \, \mathsf{USt} \equiv (\exists u)(x = (u)))$ $(\mathrm{AxIII(b1)}, \mathrm{p.} \, 45)$

(3) $(x)(x \, \beta \, \mathrm{mem_1USt} \equiv (\exists y,z)(x = \{yz\} \cdot y \, \beta \, \mathsf{USt}))$ (1), (2).

(4) $(x)(x \, \beta \, \mathrm{mem_1USt} \equiv (\exists y,z)(x = \{yz\}$
$\cdot (\exists u)(y = (u))))$ (2), (3).

In our exposition of the system we shall, however, take such moves as these as understood, and in such a case as this, proceed direct to line (4) from the axioms.

DEVELOPMENT OF THE SYSTEM: STAGE I

2.1. Stage I Theorems

In the first stage of the development of the system we aim to prove a theorem (known as the Class Theorem and to be explained later) of which essential use is made in a large number of the proofs in the later stages of the system. The following theorems, required for the proof of the Class Theorem, are all derived from AxI–AxIII.

2.11. *Immediate Consequences of AxII*

T1. *The set 0 exists.*

(The null set)

Proof

1    $(\text{vct}: b \cdot \text{vct}: c) \supset (c \subseteq b \cdot b \subseteq c)$

2    $\text{vct}: b \supset (x)(\text{vct}: x \supset x = b)$ (AxI(a), p. 45)

3    $(\exists ! \, x)\text{vct}: x$ (AxII(a), p. 45)

4   Let 0 be the set so established.

T2. *For any sets y and z, there exists a unique set s^{*zy} such that*:
$$(x)(x \, \varepsilon \, s^{*zy} \equiv x \, \varepsilon \, z \vee x = y).$$

Proof

1    $\sim(y \, \varepsilon \, z) \supset [\text{T2(II)}]$ (AxII(b), p. 45)

2    $(y \, \varepsilon \, z) \supset (x)[(x \, \varepsilon \, s \vee x = y) \equiv (x \, \varepsilon \, z)]$

3    $(y \, \varepsilon \, z) \supset (\exists ! \, s)(x)[(x \, \varepsilon \, s \vee x = y) \equiv (x \, \varepsilon \, s)]$

T3. *For any set c the set (c) exists.*

(The unit set of c)

48

Proof

(By substituting c for y, and 0 for z in T2, using T1.)

T4. *For any sets c, d, the set (cd) exists.*

(The pair set of c and d)

Proof

(By substituting c for y and (d) for z in T2, using T3.)

2.12. *Immediate Consequences of AxIII*

T5. *Every axiom in AxIII admits a unique class.*

Proof

(By AxI(b) and the definition of $\vdash\hat{\ }$, p. 40.)

T6. *The class $\wedge$ exists.*

(The empty class)

Proof

1     $\vdash\hat{\ }[a\,\beta\,[0]\cdot a\,\beta\,[(0)]]^a$ (AxIII(a1),(a3), p. 45; T3)

2   Let $\wedge$ be the class so defined.

T7. *The class St exists.*

(The class of all sets)

Proof

1     $\vdash\hat{\ }[\sim(a\,\beta\,\wedge)]^a$ (AxIII(a2), p. 45; T6)

2   Let St be the class so defined.

T8. *The class Pr exists.*

(The class of all pairs)

Proof

1     $\vdash\hat{\ }[a\,\beta\,\mathrm{mem_1St}]^a$ (AxIII(b3), p. 45; T7)

2   Let Pr be the class so defined.

T9. *The class $|A\cup B|$ exists.*

(The sum of A and B)[1]

[1] Cf. Fig. 12, p. 24.

Proof

1     $\vdash \hat{} [a \, \beta^{\, -} \, /\bar{A} \cap \bar{B}/]^a_{A,B}$ (AxIII(a2), (a3), p. 45)

2   Let $A \cup B$ be the class so defined.

T10. *The class* $\mathrm{mem_2}B$ *exists.*

(The class of pairs whose second member is in B)

Proof

1     $\vdash \hat{} [a \, \beta \, \check{} \, \mathrm{mem_1}B]^a_B$ (AxIII(b3), (c2), p. 45)

2   Let $\mathrm{mem_2}B$ be the class so defined.

T11. *The class* $\frac{A}{B}\mathrm{Pr}$ *exists.*

(The class of pairs whose first member is in A and whose second member is in B)[1]

Proof

1     $\vdash \hat{} [a \, \beta \, /\mathrm{mem_1}A \cap \mathrm{mem_2}B/]^a_{A,B}$

                        (AxIII(b3), (a3), p. 45; T10)

2   Let $\frac{A}{B}\mathrm{Pr}$ be the class so defined.

T12. *The class* $\triangle^{\check{}} A$ *exists.*

(The converse domain of A)[2]

Proof

    (By AxIII(c1), (c2).)

T13. *If* $(z)(z \, \beta \, A \supset (\exists xyu)(z = \{\{xy\}u\}))$ *then the class* $A^{\rightarrow}$ *exists.*

(The coupling-to-the-right of A)

Proof

1     $\vdash \hat{} [d \, \beta \, ^{\check{}\rightarrow \check{}\rightarrow \check{}} A]^d_A$ (AxIII(c2), (c3), p. 45)

2   Let $A^{\rightarrow}$ be the class so defined.

T14. *The class* IdPr *exists.*

(The class of pairs having identical members)

---

[1] Cf. Fig. 21, p. 31.

[2] Cf. Fig. 19, p. 29).

Proof

1   $\vdash^\wedge \left[a\,\beta \; \left/ {}^{\triangle}_{\substack{\text{USt} \\ \text{USt}}}{}^{\vee}\text{Pr} \cap \text{MPr} \right/ \right]^a$ (T11, T12)

2   $\{ad\}\,\beta \left/ {}^{\vee}_{\substack{\text{USt} \\ \text{USt}}}\text{Pr} \cap \text{MPr} \right/$ (hyp.)

3   $a,d\,\beta\;\text{USt} \cdot d\,\varepsilon\,a$ (AxIII(c2), (a3), (b2), p. 45)

4   $(\exists y)(d = (y) \cdot a = ((y)))$ (AxIII(b1), p. 45)

5   $a = ((c))$ (4)

6   $a = \{cc\}$ ((5), II.0.11, p. 45)

7   $6 \supset a\,\beta \; \left/ {}^{\triangle}_{\substack{\text{USt} \\ \text{USt}}}{}^{\vee}\text{Pr} \cap \text{MPr} \right/$ (sim.)

8   Let IdPr be the class defined in (1).

T15. $(x)(\exists Z)(x \leftrightarrow Z)$

(Every set represents a class.)

Proof

1   $\vdash^\wedge [c\,\beta \; {}^{\triangle}\!/\text{MPr} \cap \text{mem}_2[b]/]^c_b$ (AxIII, p. 45, and T10)

2   $c\,\beta \; {}^{\triangle}\!/\text{MPr} \cap \text{mem}_2[b]/$ (hyp.)

3   $(\{cd\}\,\beta\,/\text{MPr} \cap \text{mem}_2[b]/ \cdot (d = b)$ (AxIII, p. 45, and T10)

4   $c\,\varepsilon\,b$ (3), AxIII(b2), p. 45)

5   $4 \supset 2$ (sim.)

6   $b \leftrightarrow {}^{\triangle}\!/\text{MPr} \cap \text{mem}_2[b]/$ (1)–(5), dfs., p. 44)

7   $(\exists Z)(b \leftrightarrow Z)$ (6)

2.2. Standard Expressions and the Class Theorem

2.21. So far, the only expressions available as defining conditions for classes are those admitted either directly or indirectly by AxIII. But we are now able to prove that every expression which conforms to a certain standard is a defining condition for a class. This standard is specified by reference to the structure of the expression and is thus purely formal. Any expression conforming to it will be called a "standard expression".

2.22. *Definition*

Primary expressions *for* any of the following expressions:

$$a \; \varepsilon \; b \qquad\qquad a = b \qquad\qquad a \; \beta \; B$$

Note: In the remainder of this section we shall be concerned with proofs about relationships which hold between certain expressions in our symbolism—in particular with the proof that every standard expression is an authorized predicate. We shall also be investigating the way in which certain expressions can be built up out of others. In short we shall for the most part be *talking about* expressions of the system rather than actually using them. Hence in the above definition and for the remainder of this section, lower-case letters and capitals are used as symbols designating any given lower-case letters and capitals, respectively.

2.23. *Definition*

Standard expression for: Any expression which is either primary or obtainable from primary expressions by means of the constants: $\cdot$, $\vee$, $\sim$, $\supset$, (x), $(\exists x)$, (every quantifier being applied to free variables for sets only).

2.24. *The Class Theorem* (CT): *If* $[\phi]_{\gamma_1, \, \ldots, \, \gamma_n}^{v_1, \, \ldots, \, v_k}$ *is a standard expression then* $\vdash \hat{\ } [\phi]_{\gamma_1, \, \ldots, \, \gamma_n}^{v_1, \, \ldots, \, v_k}$.

Proof

The proof will be in two parts, in the first of which we prove that the assertion (CT) holds for all possible cases of primary standard expressions.

Proof: part I.

Let $[\phi]_{\gamma_1, \, \ldots, \, \gamma_n}^{v_1, \, \ldots, \, v_k}$ be a primary expression.

Then it has one of the forms:

(i) $v_1 \, \beta \, C$ (iv) $v_1 \, \varepsilon \, v_2$ (vii) $v_2 = v_1$

(ii) $v_1 = r$ (v) $v_2 \, \varepsilon \, v_1$ (viii) $v_1 \, \varepsilon \, r$

(iii) $r = v_1$ (vi) $v_1 = v_2$ (ix) $r \, \varepsilon \, v_1$

where the parameters are indicated by C and r (referring to classes and to sets, respectively). Now, we have:

(i)	In case $v_1 \, \beta \, C$	C is the required class.
(ii)	In case $v_1 = r$	The required class exists. (AxIII(a1), p. 45).
(iii)	In case $r = v_1$	sim: (ii).
(iv)	In case $v_1 \, \varepsilon \, v_2$	(AxIII(b2), (c2), p. 45).
(v)	In case $v_2 \, \varepsilon \, v_1$	sim: (iv).
(vi)	In case $v_1 = v_2$	(T14, p. 50).
(vii)	In case $v_2 = v_1$	sim: (vi).
(viii)	In case $v_1 \, \varepsilon \, r$	(T15, p. 51).
(ix)	In case $r \, \varepsilon \, v_1$	(See next step in proof, below.)

Where $r \, \varepsilon \, v_1$, the required class can be proved to exist as follows:

Sub-proof

1	$v_1 \beta \overset{\triangle \smile}{} /\mathsf{MPr} \cap \mathrm{mem}_1[r]/$	(hyp.)
2	$\{dv_1\} \beta \, /\mathsf{MPr} \cap \mathrm{mem}_1[r]/ \cdot (d = r)$	(AxIII(b3), (c1), (c2), (a3), (a1), p. 45)
3	$r \, \varepsilon \, v_1$	(AxIII(a3), (b2), p.45)
4	$3 \supset 1$	(sim.)
5	The required class is $\overset{\triangle \smile}{} /\mathsf{MPr} \cap \mathrm{mem}_1[r]/$	

This completes part I of the proof of the class theorem.

Proof: part II

We now consider all possible cases of non-primary standard expressions. But disjunction and implication are both expressible in terms of negation and conjunction, and universal quantification is expressible in terms of negation and existential quantification. It remains to be shown that the required class exists in the cases that the expression in question is constructed from a primary expression by means of either "$\cdot$" or "$\sim$" or "$\exists$".

Case 1

Let it be the case that $\vdash \hat{~}[\psi]^{v_1, \ldots, v_k}_{\gamma_1, \ldots, \gamma_m}$.

Let the class so defined be A (for the values $\alpha_1, \ldots, \alpha_m$ of $\gamma_1, \ldots, \gamma_m$).

$(v_1, \ldots, v_k)[(\psi \equiv \{v_1\{v_2 \ldots \{v_{k-1}\, v_k\}. \ldots\}\} \beta\, A) \cdot (z)(z\, \beta\, A \supset {}^kT\!: z)]$

$(v_1, \ldots, v_k)[(\sim\!\psi \equiv \{v_1\{v_2 \ldots \{v_{k-1}\, v_k\}. \ldots\}\} \beta\, \bar{A}) \cdot (z)(z\, \beta\, A \supset {}^kT\!: z)]$ (AxIII(a2), p. 45)

$\vdash \hat{~}[\sim\!\psi]^{v_1, \ldots, v_k}_{\gamma_1, \ldots, \gamma_m}$, the class so defined being $\bar{A}$ (for the values $\alpha_1, \ldots, \alpha_m$ of $\gamma_1, \ldots, \gamma_m$).

Thus the required class exists in the case of the negation of any standard expression which is already a class condition.

Case 2

Let it be the case that $\vdash \hat{~}[\psi]^{v_1, \ldots, v_k}_{\gamma_1, \ldots, \gamma_m}$, (where k > 1).

Let the class so defined be A (for the values $\alpha_1, \ldots, \alpha_m$ of $\gamma_1, \ldots, \gamma_m$).

$(v_1, \ldots, v_k)[(\psi \equiv \{v_1\{v_2 \ldots \{v_{k-1}\, v_k\}. \ldots\}\} \beta\, A) \cdot (z)(z\, \beta\, A \supset {}^kT\!: z)]$

$(v_2, \ldots, v_k)[((\exists u)\psi \equiv \{v_2\{v_3 \ldots \{v_{k-1}\, v_k\}. \ldots\}\} \beta\, {}^{\vartriangle\smallsmile}A) \cdot (z)(z\, \beta\, {}^{\vartriangle\smallsmile}A \supset {}^{k-1}T\!: z)]$ (where $u \neq v_1, \ldots, v_k$).   (T12, p. 50)

$\vdash \hat{~}[(\exists u)\psi]^{v_2, \ldots, v_k}_{\gamma_1, \ldots, \gamma_m}$, the class so defined being ${}^{\vartriangle\smallsmile}A$ (for the values $\alpha_1, \ldots, \alpha_m$ of $\gamma_1, \ldots, \gamma_m$).

Thus the required class exists in the case of the existential quantification of any standard expression which is already a class condition.

Case 3

In considering the case of a standard expression constructed by conjunction from two other standard expressions, we proceed by stages as follows:

(*a*) We assume that the component expressions have no argument in common and prove, on this assumption, that a certain class of k-tuplets exists (where there are k arguments in the given expression).

(*b*) We then prove that from such a class the required class can in any case be constructed (whether the two component expressions have any argument in common or not).

Sub-proof (*a*)

1   Let it be the case that: $\vdash^\wedge [\psi]_{\gamma_1, \, \ldots, \, \gamma_m}^{u_1, \, \ldots, \, u_j}$

　　and $\vdash^\wedge [\phi]_{\lambda_1, \, \ldots, \, \lambda_n}^{v_1, \, \ldots, \, v_k}$.

2   Let the classes so defined be B (for the values $\alpha_1, \, \ldots, \, \alpha_m$ of $\gamma_1, \, \ldots, \, \gamma_m$) and D (for the values $\omega_1, \, \ldots, \, \omega_n$ of $\lambda_1, \, \ldots, \, \lambda_n$) respectively.

3   $(u_1, \, \ldots, \, u_j)[(\psi \equiv \{u_1\{u_2 \ldots \{u_{j-1} \, u_j\} \ldots\}\} \, \beta \, B) \cdot$
　　　　　　　　　　　　　$(z)(z \, \beta \, B \supset {}^j T \colon z)]$

4   $(v_1, \, \ldots, \, v_k)[(\phi \equiv \{v_1\{v_2 \ldots \{v_{k-1} \, v_k\} \ldots\}\} \, \beta \, D) \cdot$
　　　　　　　　　　　　　$(z)(z \, \beta \, D \supset {}^k T \colon z)]$

5   $(c,e)[(c \, \beta \, B \cdot e \, \beta \, D) \equiv \{ce\} \, \beta \, \dfrac{B}{D} \mathrm{Pr}]$　　　　(T11, p. 50)

We have, as $\dfrac{B}{D}\mathrm{Pr}$, a class whose elements are the $(j + k)$-tuplets $\{ce\}$, (c being a normal j-tuplet corresponding to a system of values $c_1, \, \ldots, \, c_j$ of $u_1, \, \ldots, \, u_j$, and e being a normal k-tuplet corresponding to a system of values $e_1, \, \ldots, \, e_k$ of $v_1, \, \ldots, \, v_k$) such that: $(\psi \cdot \phi)$ holds of $c_1, \, \ldots, \, c_j, \, e_1, \, \ldots, \, e_k$. This class is not, however, a class of normal $(j+k)$-tuplets. We have now to show that the required class can be constructed from it. In doing so, we allow at the same time for the case in which the two component expressions have at least one argument in common.

The course of the argument will be as follows: Suppose that the result of rebracketing every element of $_D^B\text{Pr}$, above, in such a way that it becomes a normal $(k+j)$-tuplet, is a class. Suppose that the result of applying the same permutation to the members of each element of such a class, is a class. And suppose that the result of omitting from a class of normal $(m+1)$-tuplets all $(m+1)$-tuplets whose first two members differ, is a class. Now, by the first of these suppositions, we can obtain from $_D^B\text{Pr}$ the required class of normal $(k+j)$-tuplets. If, on the other hand, ψ and ϕ had had at least one argument in common, we could obtain, in the same way, a class of normal $(k+j)$-tuplets each of which had at least two identical members. By the second supposition, above, we could re-order each of the elements of this class so that in each case the first two members were identical. The resulting class would, by the third supposition, be a sub-class of the class of all sets of the form $\{c\{cd\}\}$. Then we should have, as the domain of the converse of this class, the class of normal m-tuplets (m $= k + j - 1$) defined by $(\psi \cdot \phi)$. This would be the required class. Thus the proofs of these three suppositions together with sub-proof (a), will constitute a proof for Case 3.

Sub-proof (b)

1	Let C be a class of normal $(m+1)$-tuplets.	
2	$m > 1$	(hyp.)
3	$a\,\beta\,/C \cap /\text{mem}_1\text{ldPr}/^{\rightarrow}/$	(hyp.)
4	$a = \{c\{cd\}\} \cdot a\,\beta\,C$	(T13, T14, p. 50)
5	$m = 1$	(hyp.)
6	$a\,\beta\,/C \cap \text{ldPr}/$	(hyp.)
7	$a = \{cc\} \cdot a\,\beta\,C$	(T14, p. 50)
8	In any case we have a class constructed from C by omitting those $(m+1)$-tuplets whose first members differ (and the domain of the converse of the class is a class of normal m-tuplets).	

Note: The remaining two suppositions (about rebracketing and permutation, respectively) will be proved simultaneously in virtue of sub-proofs c(i) and c(ii) given below. For, as sub-proof c(i) we shall prove: that a given k-tuplet can always be transformed into any prescribed normal k-tuplet having the same members, by the successive applications of a certain specified set of steps; and as sub-proof c(ii) we shall prove: that the result of applying a process of the specified kind to the members of any class of k-tuplets (to whose members such a process is applicable) is a class.

Sub-proof (c)

In the first part of this sub-proof we are simply concerned to show that four specified steps are sufficient to enable us to pass, by permutation and rebracketing, to any prescribed normal k-tuplet having the same members as some given k-tuplet. The steps in question are:

S_1^k: Replacing a k-tuplet regarded as a pair by the converse pair.

S_2^k: Coupling to the left or right applied to a k-tuplet.

S_3^k: Replacing a pair which is a member of a pair p (p being a k-tuplet) by its converse.

S_4^k: Coupling to the left or right applied to a member of a pair p (p being a k-tuplet).

LEMMA 1 (sub-proof c(i)): *For any* k-*tuplet, if c is a member of degree higher than* 1, *then by* S_1^k *and* S_2^k *the degree of c can be lowered by one.*

Proof

1   The given k-tuplet is of the form $\{p\{qr\}\}$ or $\{\{qr\}p\}$ and c is a part of $\{qr\}$.

2   Let it have the form $\{p\{qr\}\}$.

3   c is part of r or r itself or part of q or q itself.

4   c is part of r or r itself.   (hyp.)

5     $\{\{pq\}r\}$ is the result of applying S_2^k to $\{p\{qr\}\}$.

6   In $\{\{pq\}r\}$ r is lower in degree by one than in $\{p\{qr\}\}$.

7   The degree of c is lower by one.

8   c is part of q or q itself.   (hyp.)

9      $\{q\{rp\}\}$ is the result of applying S_1^k and S_2^k to $\{p\{qr\}\}$.

10   The degree of c is lower by one.   (sim: 4–6)

11   Let it have the form $\{\{qr\}p\}$.

12   The degree of c can be lowered by one using S_1^k and S_2^k.   (sim: 2–10)

From this lemma it follows immediately that the degree of any member of a k-tuplet can be lowered to 1 (provided that its degree is, to begin with, higher than 1) by successive applications of S_1^k and S_2^k.

LEMMA 2 (sub-proof c(i)): *Any k-tuplet of which c is a member* (k > 1) *can be transformed by* S_1^k *and* S_2^k *into a k-tuplet of the form* $\{bc\}$.

Proof

1   Let c be a k-tuplet member lowered to degree 1 in virtue of Lemma 1 (above).

2   The k-tuplet in question has one of the forms $\{cb\}$, $\{bc\}$.

3   The result of applying S_1^k to $\{cb\}$, is $\{bc\}$.

Sub-proof c(i)

Let p be the given k-tuplet, to be transformed into a normal k-tuplet $\{b_1\{b_2 \ldots \{b_{k-1} b_k\} \ldots \}\}$ where $b_1, \ldots, b_k$ are the members of p in some arbitrarily chosen order.

p can be transformed into a k-tuplet $\{cb_k\}$, c being a $(k-1)$-tuplet having $b_1, \ldots, b_{k-1}$ as its members.

(Lemma 2 (above))

$\{cb_k\}$ is the required normal k-tuplet in case k = 2.

k > 2.   (hyp.)

c can be transformed into a $(k-1)$-tuplet $\{db_{k-1}\}$, d being a $(k-2)$-tuplet having $b_1, \ldots, b_{k-2}$ as its members.

(Lemma 2 (above))

p can be transformed into the k-tuplet $\{\{db_{k-1}\}b_k\}$, by the application of S_3^k and S_4^k to $\{cb_k\}$.

p becomes $\{d\{b_{k-1} b_k\}\}$, by application of S_2^k, which is the required normal k-tuplet in case k = 3.

$k > 3.$ (hyp.)

d can be transformed into a $(k-2)$-tuplet $\{eb_{k-2}\}$, e being a $(k-3)$-tuplet having $b_1, \ldots, b_{k-3}$ as its members.

(Lemma 2, p. 58)

$\{d\{b_{k-1}\, b_k\}\}$ becomes $\{\{eb_{k-2}\}\{b_{k-1}\, b_k\}\}$, by application of the steps S_3^k and S_4^k.

$\{\{eb_{k-2}\}\{b_{k-1}\, b_k\}\}$ becomes $\{e\{b_{k-2}\{b_{k-1}\, b_k\}\}\}$, by application of S_2^k, which is the required normal k-tuplet in case $k = 4$.

$k > 4.$ (hyp.)

At most $k-1$ repetitions of the above steps will transform p into the required normal k-tuplet $\{b_1\{b_2 \ldots \{b_{k-1}\, b_k\} \ldots\}\}$, the steps all being of the kind S_1^k, S_2^k, S_3^k, S_4^k.

This completes the sub-proof c(i). In sub-proof c(ii) we shall prove that the result of applying any of these steps to the members of a class of k-tuplets (to whose members such a step is applicable) is a class. For this purpose we require four lemmas.

LEMMA 1 (sub-proof c(ii)): *The result of applying S_1^k or S_2^k to the members of a class of k-tuplets (to whose members such steps are applicable) is a class.*

Proof

1   Let B be a class of k-tuplets.   (hyp.)

2    $\smile B$ is the required class, for S_1^k.   (AxIII(c2), p. 45)

3    $\rightarrow B$ or $B^\rightarrow$ is the required class, for S_2^k.   (AxIII(c3), p. 45 and T13, p. 50)

It remains to be shown that the required class exists in case the steps applied are S_3^k or S_4^k. The lemma needed for this purpose is also essential to some of the proofs in the later part of the system. We shall refer to it as the Composition Lemma.

LEMMA 2 (sub-proof c(ii)): *If A and B are classes of pairs, there exists the class of all pairs $\{ab\}$ for which there exists a set x such that $\{ax\}\ \beta\ A$ and $\{xb\}\ \beta\ B$.* (The composition lemma.)

Proof

1    $d \ \beta \ /\mathrm{mem}_1 A \ \cap \ {}^\rightarrow\mathrm{mem}_2 B/$ (hyp.)

2    $d = \{\{fg\}h\} \cdot \{fg\} \ \beta \ A \cdot \{gh\} \ \beta \ B$ (AxIII(a3), (b3), (c3), p. 45), T10, p. 50

3    $c \ \beta \ \breve{/}\mathrm{mem}_1 A \ \cap \ {}^\rightarrow\mathrm{mem}_2 B/$ (hyp.)

4    $c = \{h\{fg\}\} \cdot \{fg\} \ \beta \ A \cdot \{gh\} \ \beta \ B$ (AxIII(c2), p. 45)

5    $b \ \beta \ {}^{\triangle\rightarrow\breve{}} /\mathrm{mem}_1 A \ \cap \ {}^\rightarrow\mathrm{mem}_2 B/$ (hyp.)

6    $b = \{hf\} \cdot \{fg\} \ \beta \ A \cdot \{gh\} \ \beta \ B$ (AxIII(c1), (c3), p. 45)

7    $a \ \beta \ {}^{\breve{}\triangle\rightarrow\breve{}} /\mathrm{mem}_1 A \ \cap \ {}^\rightarrow\mathrm{mem}_2 B/$ (hyp.)

8    $a = \{fh\} \cdot \{fg\} \ \beta \ A \cdot \{gh\} \ \beta \ B$ (AxIII(c2), p. 45)

9    $a = \{fh\} \cdot (\exists x)(\{fx\} \ \beta \ A \cdot \{xh\} \ \beta \ B)$ (8)

10    ${}^{\breve{}\triangle\rightarrow\breve{}} /\mathrm{mem}_1 A \ \cap \ {}^\rightarrow\mathrm{mem}_2 B/$ is the required class.

Note: The application of the composition lemma to a class of pairs *A* and a class of pairs *B*, will be called the **composition of A with B,** or $\mathrm{Cmp}\frac{A}{B}$.

LEMMA 3 (sub-proof c(ii)): *There exists the class of all sets of the form* $\{\{ab\}\{ba\}\}$.

Proof

1    $d \ \beta \ \mathrm{mem}_1\mathsf{IdPr}$ (hyp.)

2    $d = \{hg\} \cdot h = \{ff\}$ (AxIII(b3), p. 45; T14, p. 50)

3    $c \ \beta \ {}^{\rightarrow\breve{}}\mathrm{mem}_1\mathsf{IdPr}$ (hyp.)

4    $c = \{\{gf\}f\}$ (AxIII(c2), p. 45; T13, p. 50)

5    $d \ \beta \ \mathrm{mem}_1 /{}^{\rightarrow\breve{}}\mathrm{mem}_1\mathsf{IdPr}/$ (hyp.)

6    $d = \{\{\{gf\}f\}p\}$ (AxIII(b3), p. 45)

7    $k \ \beta \ /\mathrm{mem}_1/{}^{\rightarrow\breve{}}\mathrm{mem}_1\mathsf{IdPr}//{}^\rightarrow$ (hyp.)

8    $k = \{\{gf\}\{fp\}\}$ (T13, p. 50)

9    $D = /\mathrm{mem}_1/{}^{\rightarrow\breve{}}\mathrm{mem}_1\mathsf{IdPr}//{}^\rightarrow$ (hyp.)

10     $r \beta D \cap {}^{\smile}D$ (hyp.)

11     $r = \{\{gf\}\{fp\}\} \cdot r = \{\{nq\}\{sn\}\}$ (AxIII(c2), (a3), p. 45; 8)

12     $r = \{\{gf\}\{fg\}\}$ (11), FAx, p. 45; (cf. II.0.11, p. 36))

13     $D \cap {}^{\smile}D$ is the required class

LEMMA 4 (sub-proof c(ii)): *There exists the class of all sets of the form:* $\{\{g\{fk\}\}\{\{gf\}k\}\}$.

Proof

1   Let H be the class admitted by Lemma 3, p. 60.

2     $r \beta \operatorname{Cmp}_H^H$ (hyp.)

3     $r = \{\{gf\}\{gf\}\}$ (Lemma 2, p. 59)

4     $p \beta \operatorname{mem}_1 \Big/ {}^{\rightarrow}\operatorname{Cmp}_H^H \Big/$ (hyp.)

5     $p = \{\{\{\{gf\}g\}f\}k\}$ (AxIII(b3), p. 45)

6     $q \beta \Big/ \operatorname{mem}_1 \Big/ {}^{\rightarrow}\operatorname{Cmp}_H^H \Big/ \Big/ {}^{\rightarrow\rightarrow}$ (hyp.)

7     $q = \{\{gf\}\{g\{fk\}\}\}$ (T13, p. 50)

8   Let G be the class admitted by line 6. (R.I.C.)

9     $s \beta \big/ \operatorname{mem}_1 {}^{\smile}G \big/ {}^{\rightarrow}$ (hyp.)

10    $s = \{\{g\{fk\}\}\{\{gf\}n\}\}$ (AxIII(b3), p. 45)

11   Let D be the class admitted in the proof of Lemma 3, p. 58.

12    $c \beta {}^{\rightarrow\smile}\big/ \operatorname{mem}_1 D \big/$ (hyp.)

13    $c = \{\{g\{fk\}\}\{kn\}\}$ (AxIII(b3), (c2), (c3), p. 45)

14    $d \beta \operatorname{Cmp}_H^{\rightarrow\smile}\operatorname{mem}_1 D$ (hyp.)

15    $d = \{\{g\{fk\}\}\{nk\}\}$ (Lemma 2, p. 59)

16    $a \beta \Big/ \big/ \operatorname{mem}_1 {}^{\smile}G \big/ {}^{\rightarrow} \cap \operatorname{Cmp}_H^{\rightarrow\smile}\operatorname{mem}_1 D \Big/$
       (hyp.)

17     $a = \{\{g\{fk\}\{\{gf\}k\}\}$ (AxIII(a3), p. 45)

18    $\Big/ |\text{mem}_1 {}^\vee G|^{\rightarrow} \cap \text{Cmp}_H^{\rightarrow\ {}^\vee \text{mem}_1 D} \Big/$ is the required class.

Sub-proof c(ii)

Let C be a class of k-tuplets, such that S_1^k, S_2^k, S_3^k or S_4^k are applicable to its members.

The result of applying S_1^k or S_2^k to the members of C is a class. (Lemma 1.)

Let H be the class admitted by Lemma 3.

$p \, \beta \, C.$

p is of one of the forms: $\{\{ab\}c\}$, $\{a\{bc\}\}$, $\{\{ab\}\{cd\}\}$.

Let p be of the form $\{a\{bc\}\}$.

$\{a\{bc\}\} \, \beta \, C \cdot \{\{bc\}\{cb\}\} \, \beta \, H.$

$\{a\{cb\}\} \, \beta \, \text{Cmp}_H^C$

$\{a\{cb\}\}$ is the result of applying S_3^k to $\{bc\}$ in p.

In case p is of the form $\{\{ab\}\{cd\}\}$ the proof is analogous for the result of applying S_3^k to $\{cd\}$ in p.

Let p be of the form $\{\{ab\}c\}$.

$\{c\{ab\}\} \, \beta \, {}^\vee C.$

$\{c\{ba\}\} \, \beta \, \text{Cmp} \, {}^\vee{}_H^C$

$\{c\{ba\}\}$ is the result of applying S_3^k to $\{ab\}$ in p.

In case p is of the form $\{\{ab\}\{cd\}\}$ the proof is analogous for the result of applying S_3^k to $\{ab\}$ in p.

The result of applying S_3^k to those members of C to which it is applicable is the class $\Big/ \text{Cmp}_H^C \cup \text{Cmp} \, {}^\vee{}_H^C \Big/$.

Let G be the class admitted by Lemma 4.

The result of applying S_3^k to those members of G to which it is applicable is the class $\Big/ \text{Cmp}_G^C \cup \text{Cmp} \, {}^\vee{}_G^C \Big/$. The proof is analogous to the proof for the application of S_3^k.

The result of applying S_1^k, S_2^k, S_3^k or S_4^k to those members of C to which such a step is applicable, is a class.

This completes the sub-proof c(ii) and thereby the proof for case 3 (see above, page 53). As we have now considered all possible cases of non-primary standard expressions, this completes the proof of the class theorem.

2.25. *Immediate Consequences of the Class Theorem*

$CTc(i)$ $\vdash\hat{} [(x)(x \, \beta \, C \supset a \, \varepsilon \, x)]^a_C$

$CTc(ii)$ $\vdash\hat{} [(\exists x)(x \, \beta \, C \cdot a \, \varepsilon \, x)]^a_C$

$CTc(iii)$ $\vdash\hat{} [(x)(x \, \varepsilon \, a \supset x \, \beta \, C)]^a_C$

$CTc(iv)$ $\vdash\hat{} [(x)(x \, \varepsilon \, b \supset a \, \varepsilon \, x)]^a_b$

$CTc(v)$ $\vdash\hat{} [(\exists x)(x \, \varepsilon \, b \cdot a \, \varepsilon \, x)]^a_b$

$CTc(vi)$ $\vdash\hat{} [(x)(x \, \varepsilon \, a \supset x \, \varepsilon \, b)]^a_b$

The classes so defined will be referred to as follows:

By CTc(i), the intersection of the elements of the class C, or $\cap C$ (cf. Fig. 14, p. 25).

By CTc(ii), the sum of the elements of the class C, or $\cup C$ (cf. Fig. 13, p. 24).

By CTc(iii), the class of subsets of the class C, or $\subseteq *C$.

By CTc(iv), the intersection of the elements of the set b, or $\cap b$.

By CTc(v), the sum of the elements of the set b, or $\cup b$.

By CTc(vi), the class of subsets of the set b, or $\subseteq *b$.

DEVELOPMENT OF THE SYSTEM: STAGE II

3.1. Stage II Theorems

In the second stage of the development of the system we aim to prove the fundamental theorems underlying the elementary theory of ordinal numbers. For this purpose we make use not only of the class theorem and AxI–AxIII, but also of AxIV and its immediate consequences.

3.11. *Immediate Consequences of AxIV*

T16. $\sim$vct: $c \supset (\exists x)(x \,\varepsilon\, c \cdot \sim(\exists y)(y \,\varepsilon\, c \cdot y \,\varepsilon\, x))$

(Every non-empty set c has an element x such that c and x have no common element.)

Proof

(By T15 and AxIV.)

T17. $\sim(\exists x)(\text{rf}: x)$

(There is no reflexive set.)

Proof

1	$a \,\varepsilon\, a$	(hyp.)
2	$a \,\varepsilon\, (a) \cdot (\exists y)(y \,\varepsilon\, (a) \cdot y \,\varepsilon\, a)$	(T3, p. 48 (1))
3	$\sim(a \,\varepsilon\, a)$	(T16)

T18. $\sim(\exists x,y)(x \,\varepsilon\, y \cdot y \,\varepsilon\, x) \cdot \sim(\exists xyz)(x \,\varepsilon\, y \cdot y \,\varepsilon\, z \cdot z \,\varepsilon\, x)$

(There are no sets x,y, such that $x \,\varepsilon\, y$ and $y \,\varepsilon\, x$, nor such that $x \,\varepsilon\, y$ $y \,\varepsilon\, z$ and $z \,\varepsilon\, x$.)

Proof

(As for T17, with (ab), (abc) respectively, in place of (a).)

T19. $\sim$vct: $c \cdot \text{tr}: c \supset 0 \,\varepsilon\, c$

(The null set is an element of every non-empty transitive set.)

64

Proof

1	$(\exists x)(x\ \varepsilon\ c) \cdot \text{tr}: c$	(hyp.)
2	$b\ \varepsilon\ c \cdot \sim(\exists y)(y\ \varepsilon\ c \cdot y\ \varepsilon\ b)$	(T16, p. 64)
3	$b \subseteq c$	((1), (2), dfs, p. 44)
4	$\sim(\exists y)(y\ \varepsilon\ b)$	(2), (3)
5	$b = 0$	(T1, p. 48)

3.12. *Functions and One-to-one Correspondence*

By a **function** we shall mean a class of pairs such that no two elements of the class have the same set as first member. Thus if F is a function, then to every element a of the domain of F there will correspond a unique element b of the converse domain, such that $\{ab\}\ \beta\ F$; (but not necessarily vice versa). When this is the case the set b is known as **the value of the function** F **at** a (or: $F(a)$) and F is said to assign the set b to the set a.

> *Note:* $\lceil ab \rceil$, the unordered pair-class of sets a,b, exists by the Class Theorem; this is distinct from (ab), the unordered pair-set of sets a,b, whose existence is established by T4, p. 47. Also, as a corollary of T15, p. 49, we have: $(x)(\exists!\,A)(x \leftrightarrow A)$; this unique class will be called A^{*x} (the class represented by x).

Df.1. Fnc: $F\ =_{df}$
$(F \subseteq \text{Pr} \cdot (x,z)(x,z\ \beta\ F \cdot x \neq z \supset (u,w)(^{\triangle}[xz] = [uw] \supset u \neq w)))$
Df.2. FuSt: $b\ =_{df}$ Fnc: A^{*b}

By a **one-to-one correspondence** we shall mean a class of pairs such that both it and its converse class are functions. Thus if F is a one-to-one correspondence then to every element a of the domain of F there corresponds a unique element b of the converse domain such that $\{ab\}\ \beta\ F$, *and vice versa*.

Df.3. DbFnc: $F\ =_{df}$ (Fnc: $F \cdot$ Fnc: $^{\smile}F$)
Df.4. $A \approx B\ =_{df}$ $(\exists Z)(\text{DbFnc}: Z \cdot A = {}^{\triangle}Z \cdot B = {}^{\triangle\smile}Z)$

> *Note:* A *set* b and a *set* c will also be said to be one-to-one with each other if: $(\exists Z)(\text{DbFnc}: Z \cdot b \leftrightarrow {}^{\triangle}Z \cdot c \leftrightarrow {}^{\triangle\smile}Z)$. In the same way a class and a set, or a set and a class may be said to be one-to-one with each other.

3.13. *The Self-augment of a Set*

By T2 (p. 48) with $z = c$ and $y = c$ there exists a unique set s^{*cc}, for any given set c whose elements are itself and its own elements. Using Quine's term (Quine: ML, p. 246) we shall call such a set the **self-augment** of c (cf. Fig. 20, p. 30) and express it as: c'.

Df.A. c' for: s^{*cc}

3.2. Fundamental Theorems on Ordinals

3.21. *The Introduction of Ordinals*

Bernays, following Zermelo, introduces ordinal numbers into the system without referring in the definition to the concept of order. Thus the following theory of ordinals is not dependent on the theory of ordered sets.

Df.B. Or   *for*   the class determined by:

$$[\mathrm{tr}: a \cdot (x)(y)(y, x \; \varepsilon \; a \cdot y \neq x \supset x \; \varepsilon \; y \vee y \; \varepsilon \; x]^a$$

(An **ordinal** (or ordinal number) is a transitive set such that if b and c are any two different elements of it, either $b \; \varepsilon \; c$ or $c \; \varepsilon \; b$.)

3.22. *Lemmas to the Theorems on Ordinals*

T20.1. $\{(\sim\!\mathrm{vct}: C) \cdot (x)(x \; \beta \; C \supset \mathrm{tr}: x) \cdot (b \leftrightarrow \cap C)\} \supset \mathrm{tr}: b$

(If all the elements of a non-empty class are transitive, the set representing the intersection of those elements is also transitive.)

Proof

1	$(d \; \varepsilon \; a \cdot a \; \varepsilon \; b) \cdot \sim\!\mathrm{vct}: C \cdot (x)(x \; \beta \; C \supset \mathrm{tr}: x) \cdot (b \leftrightarrow \cap C)$	(hyp.)
2	$a \; \beta \cap C$	((1), dfs, p. 44, CTc(i), p. 61)
3	$k \; \beta \; C$	(hyp.)
4	$a \; \beta \; k$	(2), (3)
5	$d \; \beta \; a \cdot a \; \beta \; k \cdot \mathrm{tr}: k$	(1), (3), (4)
6	$d \; \beta \; k$	((5), dfs, p. 44)
7	$d \; \beta \cap C$	((3–6), CTc(i), p. 63)
		(See Fig. 22, p. 67)
8	$d \; \beta \; b$	((1), (7), dfs, p. 44)
9	$\mathrm{tr}: b$	((1–8), dfs, p. 44)

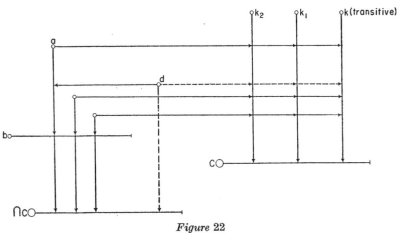

Figure 22

(Theorem 20.1, line 7)

T20.2. $\{(\sim\!\mathrm{vct}\!: C) \cdot (x)(x\,\beta\,C \supset \mathrm{tr}\!: x) \cdot (b \leftrightarrow \mathsf{U}C)\} \supset \mathrm{tr}\!: b$

(If all the elements of a non-empty class are transitive, a set representing the sum of those elements, is also transitive.)

Proof

(Analogous to the proof of T20.1.)

T20.3. $\mathrm{tr}\!: a \supset \mathrm{tr}\!: a'$

(The self-augment of a transitive set is a transitive set.)

Proof

(Immediate from df. A.)

Note: A *class* shall be said to be *transitive* $(\mathrm{tr}\!: C)$ if every element of an element is an element of it.

T20.4. $(x)(x\,\beta\,C \supset \mathrm{tr}\!: x) \supset \mathrm{tr}\!: \cap C$

(If all the elements of a class are transitive, the intersection of those elements is transitive.)

Proof

1	$(d\,\beta\,\cap C \cdot a\,\varepsilon\,d) \cdot (x)(x\,\beta\,C \supset \mathrm{tr}\!: x)$	
		(hyp.)
2	$b\,\beta\,C$	(hyp.)
3	$d\,\varepsilon\,b \cdot \mathrm{tr}\!: b$	((1), (2), dfs, p. 44, CTc(i), p. 63)
4	$a\,\varepsilon\,b$	(1), (3)
5	$a\,\beta\,\cap C$	((2–4), CTc(i), p. 63)
6	$\mathrm{tr}\!: \cap C$	((1–6), dfs, p. 44)

3.23. *Theorems on Ordinals*

T21. $(a\,\beta\,\mathsf{Or}) \supset (Z)\{(Z \subset a \cdot \mathrm{tr}\!: Z) \supset (\exists y)(y\,\varepsilon\,a \cdot y \leftrightarrow Z)\}$

(Every transitive proper sub-class of an ordinal is represented by an element of that ordinal.)

Proof

| 1 | $(a\,\beta\,\mathsf{Or}) \cdot (C \subset a \cdot \mathrm{tr}\!: C)$ | (hyp.) |
| 2 | $\vdash\hat{\ }[r\,\varepsilon\,a \cdot \sim\!(r\,\beta\,C))]^r_{a,C}$ | (CT) |

3	$(x)(x\ \beta\ D \equiv x\ \varepsilon\ a \cdot \sim(x\ \beta\ C))$	(R.I.C., p. 42)
4	$(\exists x)(x\ \varepsilon\ a \cdot \sim(x\ \beta\ C))$	(1)
5	$\sim\text{vct}: D$	((3), (4), dfs, p. 44)
6	$b\ \beta\ D \cdot \sim(\exists x)(x\ \varepsilon\ b \cdot x\ \beta\ D)$	((5), AxIV, p. 46)
7	$b\ \varepsilon\ a \cdot \text{tr}: a$	((1), (3), (6), Df.B, p. 66)
8	$k\ \varepsilon\ b$	(hyp.)
9	$k\ \varepsilon\ a$	((7), (8), dfs, p. 44)
10	$\sim(k\ \beta\ D)$	(6), (8)
11	$k\ \beta\ C$	(10), (3), (9)
		(See Fig. 23, p. 70)
12	$b \subseteq C$	((8–11), dfs, p. 44)
13	$\sim(b\ \beta\ C) \cdot \text{tr}: C$	(1), (3), (6)
14	$\sim(\exists y)(b\ \varepsilon\ y \cdot y\ \beta\ C)$	((13), dfs, p. 44)
15	$p\ \beta\ C$	(hyp.)
16	$\sim(b\ \varepsilon\ p) \cdot b \not\rightarrow p$	(13), (14), (15)
17	$p\ \varepsilon\ a \cdot a\ \beta\ \mathsf{Or}$	((1), (15), dfs, p. 44)
18	$p\ \varepsilon\ b$	((7), (16), (17), Df.B, p. 66)
19	$C \subseteq b$	((15–18), dfs, p. 44)
20	$(\exists y)(y \leftrightarrow C \cdot y\ \varepsilon\ a)$	((7), (12), (19), dfs, p. 44)

T22.1. $(a\ \beta\ \mathsf{Or}) \supset (x)\{(x \subset a \cdot \text{tr}: x) \supset x\ \varepsilon\ a\}$

(Every transitive proper sub-set of an ordinal is an element of it.)

Proof

1	$(a\ \beta\ \mathsf{Or}) \cdot (b \subset a \cdot \text{tr}: b)$	(hyp.)
2	$b \leftrightarrow C$	(T15, p. 51)
3	$(C \subseteq b \cdot b \subset a) \supset (C \subset a)$	(dfs, p. 44)
4	$k\ \varepsilon\ d \cdot d\ \beta\ C$	(hyp.)
5	$d\ \varepsilon\ b \cdot \text{tr}: b$	((1), (2), dfs, p. 44)
6	$k\ \varepsilon\ b \cdot k\ \beta\ C$	((4), (5), (2), dfs, p. 44)

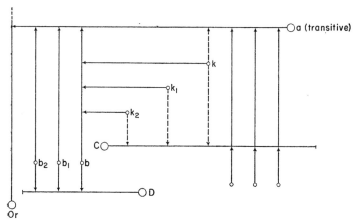

Figure 23

(Theorem 21, line 11)

7	$\mathrm{tr}: C \cdot C \subset a$	$((4\text{–}6), (1\text{–}3), \mathrm{dfs}, \mathrm{p.}\ 44)$
		(See Fig. 24, p. 72)
8	$p \leftrightarrow C \cdot p\ \varepsilon\ a$	$((1), \mathrm{T}21, \mathrm{p.}\ 68)$
9	$p = b$	$(\mathrm{AxI(a)}, \mathrm{p.}\ 45, (2), (2),$
		$(8), \mathrm{dfs}, \mathrm{p.}\ 44)$
10	$b\ \varepsilon\ a$	$(8), (9)$

T22.2. $\{(a,b\ \beta\ \mathsf{Or}) \cdot (a \subset b)\} \supset (a\ \varepsilon\ b)$

(If a,b are ordinals and a is a proper sub-set of b, then a is an element of b.)

Proof

(Immediate from T22.1 and Df.B.)

T23. $(\sim\!\mathrm{vct}: C \cdot C \subseteq \mathsf{Or}) \supset (\exists x)(x \leftrightarrow \cap C \cdot x\ \beta\ C)$

(The intersection of the elements of a non-empty class of ordinals is represented by a set belonging to that class.)

Proof

1	$\sim\!\mathrm{vct}: C \cdot C \subseteq \mathsf{Or}$	(hyp.)
2	$\mathrm{tr}: \cap C$	(T20.4, p. 68; Df.B, p. 66)
3	$a\ \beta\ C$	$((1), \mathrm{dfs}, \mathrm{p.}\ 44)$
4	$\cap C \subseteq a$	$(\mathrm{CTc(i)}, \mathrm{p.}\ 63, (3))$
5	$a \subseteq \cap C \supset (\exists x)(x \leftrightarrow \cap C \cdot x\ \beta\ C)$	$(3), (4)$
6	$\sim(a \leftrightarrow \cap C)$	(hyp.)
7	$\cap C \subset a \cdot \mathrm{tr}: \cap C$	$((2), (4), (6), \mathrm{dfs}, \mathrm{p.}\ 44)$
8	$(\exists x)(x\ \varepsilon\ a \cdot x \leftrightarrow \cap C)$	$(\mathrm{T}21, \mathrm{p.}\ 68\ (1), \mathrm{dfs}, \mathrm{p.}\ 44; (2), (7))$
9	$\cap C \leftrightarrow d$	(8)
10	$\sim(d\ \beta\ C)$	(hyp.)
11	$k\ \beta\ C$	(hyp.)
12	$\cap C \subseteq k$	$(\mathrm{CTc(i)}, \mathrm{p.}\ 63, (11))$
13	$d \subseteq k$	$((9), (12), \mathrm{dfs}, \mathrm{p.}\ 44)$
14	$k \subseteq d \supset d\ \beta\ C$	$((13), (11), \mathrm{AxI(a)}, \mathrm{p.}\ 45)$

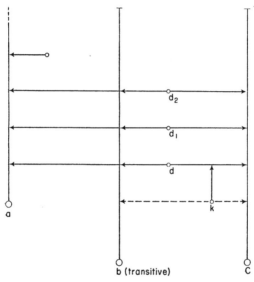

Figure 24

(Theorem 22.1, line 7)

15      $d \subset k \cdot \mathrm{tr} : d \cdot k \, \beta \, \mathsf{Or}$ ((1), (2), (9), (11), (13),
                                                                                (14), dfs, p. 44)

16      $d \, \varepsilon \, k$ (T22.1, p. 69; (15))

17      $d \, \beta \cap C$ ((11–16), CTc(i), p. 63)

18      $d \, \varepsilon \, d$ ((17), (9), dfs, p. 44)

19      $(\exists x)(x \leftrightarrow \cap C \cdot x \, \beta \, C) \cdot d \, \beta \, C \cdot \cap C \leftrightarrow d$

                                                                ((9), (10–18), T17, p. 64)

20      $\sim(a \leftrightarrow \cap C) \supset (\exists x)(x \leftrightarrow \cap C \cdot x \, \beta \, C)$

                                                                (6–19)

21      $(\exists x)(x \leftrightarrow \cap C \cdot x \, \beta \, C)$ ((5), (20), dfs, p. 44)

T24.1. $(\sim\mathrm{vct} : c \cdot c \subseteq \mathsf{Or} \cdot d \leftrightarrow \cap c) \supset (d \, \varepsilon \, c)$

(Every set representing the intersection of the elements of a non-empty set c whose elements are ordinals, is itself an element of c.)

Proof

(By T15 and T23)

T24.2. $(a,b \, \beta \, \mathsf{Or} \cdot a \neq b) \supset (a \subset b \vee b \subset a)$

(If a,b are distinct ordinals then either a is a proper sub-set of b, or vice versa.)

Proof

1      $a,b \, \beta \, \mathsf{Or} \cdot a \neq b$ (hyp.)

2      $(ab) \leftrightarrow [ab]$ (T4, p. 49; dfs, p. 44).[1]

3      $k \leftrightarrow [ab] \cdot (k = a \vee k = b)$ ((1), T23, p. 71).

4      $\cap[ab] = \cap(ab)$ (CTc(i), p. 63)

5      $k \leftrightarrow \cap(ab)$ (3), (4)

6      $k = a$ (hyp.)

7      $(c \, \varepsilon \, a) \supset (c \, \beta \cap(ab))$ ((5), (6), dfs, p. 44)

8      $b \, \varepsilon \, (ab)$ (T4, p. 49)

[1] *Or:* $(ab) \leftrightarrow P$
           $k \leftrightarrow \cap P$
           $\cap P = \cap(ab)$
              (etc., with T24.1)

9     $c \, \varepsilon \, a \supset c \, \varepsilon \, b$ ((7), (8), CTc(i), p. 63)

10    $a \subset b$ ((9), (1), dfs, p. 44)

11    $(k = b) \supset (b \subset a)$ (sim. 6–10).

12    $a \subset b \lor b \subset a$ (3), (6–10), (11)

Note: The relation of *proper sub-class of* between ordinals constitutes what is known as an ordering relation. Thus we can speak of an ordinal b being lower than another ordinal c in case $b \subset c$, and vice versa. It may be found helpful if, subsequently, theorems are read in this sense, wherever the sub-class relation is asserted between ordinals. For example, T22.2 asserts that the lower of two distinct ordinals is always an element of the higher.

T24.3. $(\mathrm{tr}\colon c \cdot c \subseteq \mathsf{Or}) \supset c \, \beta \, \mathsf{Or}$

(Every transitive set of ordinals is an ordinal.)

Proof

(By T22.2, T24.2 and Df.B).

T24.4. $(\sim\!\mathrm{vct}\colon C \cdot C \subseteq \mathsf{Or}) \supset (\exists x)\{x \, \beta \, C \cdot (y)(y \, \beta \, C \cdot y \neq x \supset x \subset y)\}$

(Among the elements of a non-empty class of ordinals, there is always a lowest one.)

Proof

1    $\sim\!\mathrm{vct}\colon C \cdot C \subseteq \mathsf{Or}$ (hyp.)

2    $s \leftrightarrow \cap C \cdot s \, \beta \, C$ ((1), T23, p. 71)

3    $s \, \beta \, \mathsf{Or}$ ((1), (2), dfs, p. 44)

4    $b \, \beta \, C \cdot b \neq s$ (hyp.)

5    $s \subseteq \cap C$ ((2), dfs, p. 44)

6    $s \subset b$ ((4), (5), AxI(a) and dfs, p. 44)

7    $(\exists x)\{x \, \beta \, C \cdot (y)(y \, \beta \, C \cdot y \neq x \supset x \subset y)\}$ (2), (4–6)

T25. $(n \, \beta \, \mathsf{Or} \cdot c \, \varepsilon \, n) \supset c \, \beta \, \mathsf{Or}$

(Every element of an ordinal is an ordinal.)

Proof

1    $n \, \beta \, \mathsf{Or} \cdot c \, \varepsilon \, n$ (hyp.)

2	$c \subseteq n$	((1) Df.B, p. 66; dfs, p. 44)
3	$(a,b \, \varepsilon \, c \cdot a \neq b) \supset (a \, \varepsilon \, b \vee b \, \varepsilon \, a)$	((1), (2), Df.B, p. 66)
4	$a \, \varepsilon \, b \cdot b \, \varepsilon \, c$	(hyp.)
5	$a \neq c$	(T18, p. 64; (4))
6	$\sim c \, \varepsilon \, a)$	(T18, p. 64; (4))
7	$a,c \, \varepsilon \, n$	((1), (2), (4), Df.B., p. 66)
8	$(a \, \varepsilon \, c \vee c \, \varepsilon \, a)$	((1), (7), Df.B, p. 66)
9	$a \, \varepsilon \, c$	(6), (8)
10	tr: c	((4–10), dfs, p. 44) (See Fig. 25, p. 76)
11	$c \, \beta \, \mathrm{Or}$	((3), (10), Df.B, p. 66)

Df.C. L_n *for* the class determined by: $[a,n \, \beta \, \mathrm{Or} \cdot a \subset n]_n^a$

(The class of ordinals lower than n.)

T26.1. $n \, (\beta \, \mathrm{Or}) \supset (n \leftrightarrow L_n)$

(Every ordinal represents the class of ordinals lower than itself.)

Proof

1	$n \, \beta \, \mathrm{Or}$	(hyp.)
2	$a \, \varepsilon \, n$	(hyp.)
3	$a \, \beta \, \mathrm{Or}$	(T25, (1), (2))
4	$a \subseteq n$	((1), (2), Df.B, p. 66)
5	$a \subset n \cdot a \, \beta \, \mathrm{Or}$	((2), (4), T17, p. 64; AxI(a), p. 45)
6	$a \, \beta \, L_n$	((5), Df.C)
7	$a \, \beta \, L_n \supset a \, \varepsilon \, n$	(T22.2, p. 71; Df.C)
8	$n \leftrightarrow L_n$	((2–6), (7), dfs, p. 44)

T26.2. $(a \, \beta \, \mathrm{Or}) \supset (a' \, \beta \, \mathrm{Or})$

(The self-augment of an ordinal is an ordinal.)

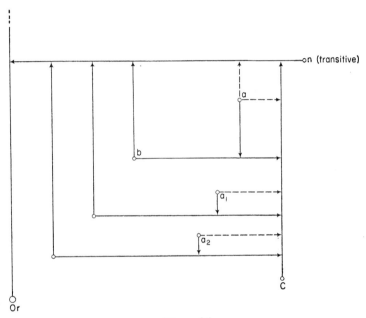

Figure 25

(Theorem 25, line 10)

Proof

(By T20.3, T25, and T24.3)

T26.3. $(C \subseteq \text{Or} \cdot a \leftrightarrow \text{U}C) \supset (a \, \beta \, \text{Or})$

(If a set represents the sum of the elements of a class of ordinals, it is itself an ordinal.)

Proof

(By T25, T24.3, and Df.B)

T26.4. $(x)(x \, \beta \, \text{Or} \cdot x \neq 0 \supset 0 \subset x)$

(0 is the lowest ordinal.)

Proof

(By T24, T22.2)

Df.5. $N_b^a \; =df \; (a,b \, \beta \, \text{Or} \cdot b \subset a \cdot \sim(\exists y)(b \subset y \cdot y \subset a))$

(a is next higher (ordinal) to b.)

T26.5. $(a \, \beta \, \text{Or}) \supset \; N_a^{a'}$

(If a is an ordinal, the next higher ordinal to a is the self-augment of a.)

Proof

1	$a \, \beta \, \text{Or}$	(hyp.)
2	$a' \, \beta \, \text{Or} \cdot a \subset a'$	(T26.2, Df.A, Df.B, p. 66)
3	$a \subset d \cdot d \subset a'$	(hyp.)
4	$b \, \varepsilon \, d \cdot \sim(b \, \varepsilon \, a)$	((3), dfs, p. 44)
5	$b \, \varepsilon \, a' \cdot \sim(b \, \varepsilon \, a)$	((3), (4), dfs, p. 44)
6	$b \neq a \supset b \, \varepsilon \, a$	(Df.A, p. 66)
7	$b = a$	(5), (6)
8	$a \subset d \cdot a \, \varepsilon \, d$	(3), (4), (7)
9	$k \, \varepsilon \, d \supset k \, \varepsilon \, a'$	((3), dfs, p. 44)
10	$k \, \varepsilon \, a \vee k = a \supset k \, \varepsilon \, d$	((8), dfs, p. 44)
11	$d = a'$	((9), (10), Df.A, p. 66)
12	$d \subset d$	(3), (11)
13	$\sim(\exists y)(a \subset y \cdot y \subset a')$	((3–12), T17, p. 64)
14	$N_a^{a'}$	((1), (2), (12), Df.5)

T26.6. $(a,b\ \beta\ \text{Or} \cdot a' = b') \supset (a = b)$

(If a,b are ordinals and the self-augment of a is identical with the self-augment of b, then a is identical with b.)

Proof

(By T18 and Df.A.)

T26.7. $\{a'\ \beta\ \text{Or} \cdot c\ \varepsilon\ a' \cdot (y)(y\ \varepsilon\ a' \cdot y \neq c \supset y \subset c)\} \supset (c = a)$

(An ordinal a' has a as its highest element.)

Proof

1	$a'\ \beta\ \text{Or} \cdot c\ \varepsilon\ a' \cdot (y)(y\ \varepsilon\ a' \cdot y \neq c \supset y \subset c)$	
		(hyp.)
2	$c \neq a$	(hyp.)
3	$a \subset c \cdot c\ \beta\ \text{Or}$	((1), (2), T25, p. 74)
4	$a\ \varepsilon\ a'$	(Df.A, p. 66)
5	$a\ \beta\ \text{Or}$	((4), (1), T25, p. 74)
6	$a\ \varepsilon\ c$	((3), (5), T22.2, p. 71)
7	$k\ \varepsilon\ c \supset k\ \varepsilon\ a'$	((1), Df.A, p. 66)
8	$(k\ \varepsilon\ a \vee k = a) \supset (k\ \varepsilon\ c)$	((3), (6), dfs, p. 44)
9	$a' = c$	((7), (8), Df.A, p. 66; AxI(a), p. 45)
10	$c = a$	((1), (9), T17, p. 64)

T26.8. $\{a\ \beta\ \text{Or} \cdot c\ \varepsilon\ a \cdot (y)(y\ \varepsilon\ a \cdot y \neq c \supset y \subset c)\} \supset (a = c')$

(An ordinal having a as its highest element is identical with a'.)

Proof

(Similar to T26.7, using also T26.2, T24.2, T18)

Note: Using AxIV and T25, it can be proved that the Class **Or** is identical with the class of transitive sets such that every member of each such set is transitive. In other words an ordinal is a transitive set such that all its elements are transitive. The force of AxIV is here apparent, for without it the unit set of any reflexive set would be an ordinal, in which case **Or** would not be the class we require.

When any two ordinals, a, c, satisfy the left-hand side of T26.8 we shall say that a has c as highest element. We therefore introduce the following definition:

Df.6. H_c^a $=df$ $\{a \ \beta \ \text{Or} \cdot c \ \varepsilon \ a \cdot (y)(y \ \varepsilon \ a \cdot y \neq c \supset y \subset c)\}$

In the remaining sections (4 and 5) strictly complete annotations for proofs will be dispensed with, sufficient indication being given for the reader to complete the annotations himself.

THE THEORY OF FINITE ORDINALS

Number theory (in particular the theory of finite ordinal numbers) is derived from Stage II in the development of the system, by defining *finite ordinal number* in terms of *ordinal*. For, with finite ordinal so defined, the Peano axioms are available as theorems of the system. All that is then required is a set-theoretic basis for the usual recursive definitions of numerical functions. This is provided by the existence theorem proved in this section. The system then has the elementary laws of arithmetic as theorems.

4.1. Finite Ordinal Number

By a **finite ordinal** is meant an ordinal such that both it and every ordinal lower than it are either 0 or of the form c'. We therefore introduce the following definition:

Df.D. FOr *for* the class determined by: $[(n = 0 \text{ v } (\exists y)(n = y'))$
$\cdot (z)(z \text{ } \varepsilon \text{ } n \supset z = 0 \text{ v } (\exists y)(z = y')) \cdot (n \text{ } \beta \text{ Or})]$

4.11. *Corollaries of the Definition of Finite Ordinal*

T30.1. $n \text{ } \beta \text{ FOr} \supset \{(\text{vct}: n \text{ v } (\exists y)\text{H}_y^n) \cdot (z)(z \subset n \supset \text{vct}: z \text{ v } (\exists y)\text{H}_y^z)\}$

(Every finite ordinal is such that it and every element of it has a highest element or no elements.)

Proof

1	$n \text{ } \beta \text{ FOr}$	(hyp.)
2	$n = 0 \supset \text{vct}: n$	(T1, p. 48)
3	$(\exists y)(n = y')$	(hyp.)
4	$n = k'$	(3)
5	$\text{H}_k^{k'}$	(T26.6, p. 78)
6	$(\exists y)\text{H}_y^n$	(5)
7	$(z)(z \text{ } \varepsilon \text{ } n \supset \text{vct}: z \text{ v } (\exists y)\text{H}_y^z)$	(sim.)

8    $\{(\text{vct}: n \vee (\exists y)\text{H}_y^n) \cdot (z)(z \, \varepsilon \, n \supset \text{vct}: z \vee (\exists y)\text{H}_y^z)\}$

$(1), (2), 3\text{--}6), (7)$

Note: By T30.1 a finite ordinal either has a highest element or has no element; and so does every element of a finite ordinal.

T30.2. $(x)(n \, \beta \, \text{FOr} \cdot x \, \beta \, L_n \supset x \, \beta \, \text{FOr})$

(Every ordinal which is lower than a finite ordinal is finite.)

Proof

(Immediate from df.D.)

4.2. The Peano Axioms as Theorems of the System

T31. $0 \, \beta \, \text{FOr}$

(0 is a finite ordinal.)

Proof

(Immediate from Df.D.)

T32. $(n \, \beta \, \text{FOr}) \supset (n' \, \beta \, \text{FOr} \cdot n' \neq 0)$

(If n is a finite ordinal then the self-augment of n is a finite ordinal and is distinct from 0.)

Proof

(Immediate from Df.D. and Df.A.)

T33. $(k,n \, \beta \, \text{FOr} \cdot k' = n') \supset (k = n)$

(If k,n are finite ordinals whose self-augments are distinct, then they are themselves distinct.)

Proof

(By T26.6 and Df.D.)

T34. $\{0 \, \beta \, C \cdot (x)(x \, \beta \, C \cdot x \, \beta \, \text{FOr} \supset x' \, \beta \, C)\} \supset (\text{FOr} \subseteq C)$

(If C is a class to which the null set belongs, and if for every finite ordinal x belonging to C, the self-augment of x belongs to C, then every finite ordinal belongs to C (Principle of Mathematical Induction).)

Proof

1    $0 \, \beta \, C \cdot (x)(x \, \beta \, C \cdot x \, \beta \, \text{FOr} \supset x' \, \beta \, C)$

(hyp.)

2	$(\exists x)(x\,\beta\,\text{FOr} \cdot \sim(x\,\beta\,C))$	(hyp.)
3	$\vdash\!\hat{}[k\,\beta\,\text{FOr} \cdot \sim(k\,\beta\,C)]$	(CT)
4	$(x)(x\,\beta\,A \equiv x\,\beta\,\text{FOr} \cdot \sim(x\,\beta\,C))$	((3), RIC)
5	$\sim\text{vct}\!: A$	(2), (4)
6	$d\,\beta\,A \cdot (y)(y\,\beta\,A \cdot y \neq d \supset d \subset y)$	(T24.4, p. 74)
7	$\sim(0\,\beta\,A)$	(1), (4)
8	$d \neq 0$	((6), (7), FAx, p. 45)
9	$(\exists s)(d = s')$	(hyp.)
10	$s\,\varepsilon\,s' \cdot s\,\beta\,\text{FOr}$	(T25, p. 74)
11	$s \subset s'$	(Df.A., p. 66)
12	$s\,\beta\,C \supset d\,\beta\,C$	(9), (10), (1)
13	$\sim(s\,\beta\,C)$	(4), (6), (12)
14	$s\,\beta\,A \cdot d \subset s$	(10), (13), (6)
15	$s' \subset s$	(9), (14)
16	$\sim(\exists x)(x\,\beta\,\text{FOr} \cdot \sim(x\,\beta\,C))$	(2–15)
17	$\text{FOr} \subseteq C$	(16)

Note: We shall write "1" for "0'", "2" for "1'" and so on. Thus the Number One is the self-augment of 0, etc.

4.3. Iteration

The theorem referred to in the first paragraph of this section (p. 80) is concerned with the existence of a particular type of function. To prove its existence we have first of all to construct a slightly less complicated function. The less complicated function is one defined by reference to some given class, and any given function which assigns to every member of that class, a member of that class. In addition, it has the finite ordinals as its domain. Thus if, for example, we had a class $A = [a,b,c]$ and a function $F = [\{ac\},\{ba\},\{cb\}]$, the function which we have first to define would be one which assigned to 0 some given member a of A, and to n' the value of F at the value of the function at n (see § 3.12, p. 65); in other words the class:

$$[\{0a\},\{1c\},\{2b\},\{3a\},\{4c\}, \ldots]$$

Such a function will be called **the iterator of F on a,** or $F^{:a}$.

The existence theorem which we require is the assertion that there exists, for every finite ordinal n, a function whose domain is the finite ordinals, such that if k is the first member of an element, the second member of that element is the value of the iterator of F on k at n; or $F^{:k}(n)$. Such a function will be called **the n-fold iteration of F**, or $\overset{n}{F}$. For example, if $A =$ FOr and F is the class:

$$[\{01\},\{10\},\{22\},\{30\},\{43\},\{50\},\{64\}, \ldots]$$

then the iterators of F on 0,1,2,3,4, respectively, are:

$$F^{:0} = [\{00\},\{11\},\{20\},\{31\},\{40\}, \ldots]$$
$$F^{:1} = [\{01\},\{10\},\{21\},\{30\},\{41\}, \ldots]$$
$$F^{:2} = [\{02\},\{12\},\{22\},\{32\},\{42\}, \ldots]$$
$$F^{:3} = [\{03\},\{10\},\{21\},\{30\},\{41\}, \ldots]$$
$$F^{:4} = [\{04\},\{13\},\{20\},\{31\},\{40\}, \ldots]$$

and the 3-fold iteration of F is:

$$\overset{3}{F} = [\{01\},\{10\},\{22\},\{30\},\{41\}, \ldots]$$

and, for the value of the 3-fold iteration of F at 4, for example, we have:

$$\overset{3}{F}(4) = 1$$

The existence assertion required, then, is that for every finite ordinal n, given a function F which assigns to every element of FOr an element of FOr, there exists the function $\overset{n}{F}$. In order to prove this we have first to prove that, for every finite ordinal n, there exists the function $F^{:n}$, for a given function F which assigns to every member of a given class A a member of A. This will be called the *Iteration Theorem* (or, IT).

4.31. *The Iteration Theorem* (IT)

   Let c be a set, A a class and $c \, \beta \, A$, and F a function which assigns to every element of A an element of A. Then there exists a function H (depending on parameters c and F)

which assigns to every finite ordinal an element of A, and satisfies the conditions:

$$H(0) = c$$
$$H(n') = F(H(n))$$

for every finite ordinal n.

The proof of IT will proceed as follows. We shall first prove the existence of a unique functional set, s_n, representing a function $H*$ whose domain is $/L_n \cup [n]/$, for every finite ordinal n; where $H*$ is otherwise exactly like H, except that in place of the condition $H(n') = F(H(n))$, we have:

$$(k)(k \; \varepsilon \; n \supset H*(k') = F(H*(k)))$$

We next prove the existence of a class of pairs $\{nb\}$ such that n is a finite ordinal and b is the value assigned to n by the unique functional set (i.e. the function represented by the functional set) satisfying the conditions just stated. We then show that this class is a function whose value for every finite ordinal is an element of A, and which satisfies the conditions stated in IT. In short we show that $F^{:c}$ exists for any element c of any given class A. Finally, it should be noticed that the class of pairs $\{nb\}$ whose existence is to be proved, can be proved to exist only by means of the Class Theorem. But an expression involving quantification over a class is not a standard expression. Hence the necessity for bringing in a functional set to represent the function $H*$. In doing so, we shall refer to the set representing the domain of a function F, as the domain of the set representing the function. Thus where we have: Fnc: F and $s \leftrightarrow F$ and $c \leftrightarrow {}^\vartriangle F$, we shall refer to c as: ${}^\vartriangle s$ and we shall use ${}^{\vartriangle\smile} s$ analogously.

Proof: part I

1   Let c,A,F, satisfy the hypothesis of IT.

2      $\vdash \widehat{}[n \; \beta \; \textbf{FOr} \cdot (\exists \, !z)\{(\text{FuSt}: z \cdot {}^\vartriangle z \leftrightarrow /\text{L}_n \cup [n]/ \cdot \{0c\} \; \varepsilon \; z \cdot (x,y)(\{xy\} \; \varepsilon \; z \cdot x \; \varepsilon \; n \supset \{x',F(y)\} \; \varepsilon \; z)]_{F,c}^n$

                                           (CT, Df. 2, p. 65; Df. 5, p. 77)

3   Let D be the class so defined.

4     $0 \, \beta \, \text{FOr} \cdot \text{FuSt}:(\{0c\}) \cdot {}^{\triangle}(\{0c\}) \leftrightarrow /L_0 \cup [0]/$
      (Df. 1, Df. 2, p. 65)

5     $\{0c\} \, \varepsilon \, (\{0c\}) \cdot (x)(\text{FuSt}: x \cdot \ldots x \ldots \supset x = (\{0c\}))$

6     $\sim(\exists x,y)(\{xy\} \, \varepsilon \, (\{0c\}) \cdot x \, \varepsilon \, 0)$

7     $0 \, \beta \, D$ (3)–(6)

8   Let n be an element of D, and s_n the required
      unique functional set.

9   Let r be $/s_n \cup (\{n'F(q)\})/$, where q is the set assigned
      by (the function represented by) s_n to n
      (AxII(b), p. 45)

10    $\{0c\} \, \varepsilon \, r \cdot n' \, \varepsilon \, {}^{\triangle}r$ (2)

11    $\{kc\} \, \varepsilon \, r \cdot k \, \varepsilon \, n'$ (hyp.)

12    $\{kc\} \, \varepsilon \, s_n \vee \{kc\} = \{n'F(q)\}$ ((9) CTc(ii), p. 63)

13    $\{kc\} \, \varepsilon \, s_n$ ((11), (12), T17, p. 64)

14    $k \, \varepsilon \, {}^{\triangle}s_n \supset k \, \beta \, /L_n \cup (n)/$ (2)

15    $k \subset n$ (hyp.)

16    $k \, \varepsilon \, n \cdot \{k'F(c)\} \, \varepsilon \, s_n$ (T22.2, p. 71)

17    $\{k'F(c)\} \, \varepsilon \, r$ (9)

18    $k = n$ (hyp.)

19    $\{kc\} = \{nc\}$ (18)

20    $c = q \cdot k' = n'$ (9), (18)

21    $\{k'F(c)\} = \{n'F(q)\}$ (20)

22    $k \subset n \vee k = n$ ((13), (14), Df.C, p. 75)

23    $(x,y)(\{xy\} \, \varepsilon \, r \cdot x \, \varepsilon \, n' \supset \{x'F(y)\} \, \varepsilon \, r)$ (11–12), (15–17), (18-21)

24    $n \, \beta \, D \supset n' \, \beta \, D$ ((8–23), T32, p. 81)

25    $\text{FOr} \subseteq D$ (T34, p. 81)

26    $(x)(x \, \beta \, D \supset x \, \beta \, \text{FOr})$ (2)

27    $D = \text{FOr}$ (25), (26)

Proof: part II

1 Let D be the class admitted in part I of the proof.

2 $\vdash \hat{\ } [n\,\beta\,D \cdot \{nb\}\,\varepsilon\,s_n]^{n,b}_{F,c}$ (s_n being the functional set for n, referred to in the defining condition for D)   (CT)

3 Let H be the class so defined.

4 $H \subseteq \mathsf{Pr}$ (T8, p. 49)

5 $\{pq\},\{fg\}\,\beta\,H \cdot \{pq\} \neq \{fg\}$ (hyp.)

6 $p = f$ (hyp.)

7 $\{fq\},\{fg\}\,\varepsilon\,s_n \cdot \mathsf{FuSt}{:}\,s_n$ (5), (2)

8 $\{fq\} = \{fg\} \cdot q = g \cdot q \neq g$ ((6), Df. 1, Df. 2, p. 65)

9 $p \neq f$ (6)–(8)

10 $\mathsf{Fnc}{:}\,H$ ((5)–(9), Df. 1, p. 65)

11 $^{\triangle}H = \mathsf{FOr}$ (2)

12 $(n\,\beta\,\mathsf{FOr} \cdot k = H(n)) \supset \{nk\}\,\varepsilon\,s_n$ (2), (3)

13 $k = H(0)$ (hyp.)

14 $\{0k\}\,\varepsilon\,s_n \cdot \mathsf{FuSt}{:}\,s_n$ (2), (3)

15 $\{0k\} = \{0c\} \cdot k = c$ ((14), Df. 2, p. 65)

16 $H(0) = c$ (13), (15)

17 $\mathsf{FuSt}{:}\,r \cdot {}^{\triangle}r \leftrightarrow /\mathsf{L}_{n'}\,\cup\,(n')/ \cdot \{0c\}\,\varepsilon\,r \cdot (x,y)(\{xy\}\,\varepsilon\,r \cdot x\,\varepsilon\,n'$
$\supset \{x'F(y)\}\,\varepsilon\,r)$ (hyp.)

18 $\{nk\}\,\varepsilon\,r \cdot \{nd\}\,\varepsilon\,s_n$ (hyp.)

19 $k \neq d$ (hyp.)

20 $s_n \neq r \cdot \{0c\}\,\varepsilon\,r \cdot \{0c\}\,\varepsilon\,s_n$ (FAx, p. 45)

21 $s_n \subset r \vee r \subset s_n$ (20)

22 $r \subset s_n \supset \{nk\}\,\varepsilon\,s_n$ (18)

23 $r \subset s_n \supset n \neq n$ (18), (22)

24 $s_n \subset r \supset \{nd\}\,\varepsilon\,r$ (18), (19)

25 $s_n \subset r \supset n \neq n$ (18)

26 $k = d$ (19)–(25)

27 $(\{nk\}\,\varepsilon\,r \cdot \{nd\}\,\varepsilon\,s_n) \supset k = d$ (18)–(26)

28	$g = H(n') \cdot b = H(n)$	(hyp.)
29	$n \, \varepsilon \, {}^{\triangle}r$	(17)
30	$(\exists y)(\{ny\} \, \varepsilon \, r)$	(§4.31, p. 83)
31	$(\{na\} \, \varepsilon \, r \cdot \{nb\} \, \varepsilon \, s_n) \supset a = b$	(27)
32	$\{nb\} \, \varepsilon \, r$	(30), (31)
33	$\{n'F(b)\} \, \varepsilon \, r$	((17), (32), Df.A, p. 66)
34	$\{n'g\} \, \varepsilon \, r$	(17), (28)
35	FuSt: r	(17)
36	$g = F(b)$	(33)–(35)
37	$H(n') = F(H(n))$	(17)–(36)
38	$H(0) \, \beta \, A$	(16)
39	$H(n) \, \beta \, A$	(hyp.)
40	$\{H(n)q\} \, \beta \, F \cdot q \, \beta \, A \cdot H(n') = q$	(37)
41	$H(n') \, \beta \, A$	(2), (40)
42	$(n)(H(n) \, \beta \, A) \cdot H = F^{:c}$	((38), (39–41), T34, p. 81)

This completes the proof of the iteration theorem (cf. lines (10), (11), (16), (37), and (42)), since H is the function $F^{:c}$ (see §4.3, p. 82).

4.4. The Existence Theorem for the Function: $\overset{n}{F}$. (ET)

Let c be a set, A a class and c β A and F a function which assigns to every element of A an element of A. Then there exists, for every finite ordinal n, a function which assigns to every element b of A, the set $F^{:b}(n)$.

Proof

1	Let c, A, F satisfy the hypothesis of ET.	
2	$\vdash \hat{} [x \, \beta \, A \cdot y = F^{:x}(n)]_n^{xy}$	(CT, IT)
3	Let the class so defined be $\overset{n}{G}$.	
4	$p,q \, \beta \, \overset{n}{G} \cdot p = \{bd\} \cdot q = \{fk\} \cdot p \neq q$	(hyp.)
5	$b = f$	(hyp.)

6	$d = F^{:b}(n) \cdot k = F^{:f}(n)$	(2), (4)
7	$d = k$	(5), (6)
8	$b \neq f$	(4), (5–7)
9	$\text{Fnc}: \overset{n}{G}$	((4–8), Df. 1, p. 65)
10	$\vdash \hat{\ } [n \ \beta \ \mathsf{FOr} \cdot (\exists z)(z \ \beta \ \overset{n}{G})]^n$	(CT)
11	Let the class so defined be K.	
12	$c \ \beta \ A \cdot c = F^{:c}(0)$	(IT)
13	$. \{cc\} \ \beta \ \overset{0}{G}$	(12), (2)
14	$0 \ \beta \ K$	(13)
15	$n \ \beta \ K$	(hyp.)
16	$\{xy\} \ \beta \ \overset{n}{G}$	(15)
17	$x \ \beta \ A \cdot y = F^{:x}(n)$	(16), (2)
18	$(\exists z)(z = F^{:x}(n'))$	(IT)
19	$d = F^{:x}(n')$	(18)
20	$x \ \beta \ A \cdot d = F^{:x}(n')$	(17), (19)
21	$\{xd\} \ \beta \ \overset{n'}{G}$	(20), (2)
22	$n' \ \beta \ K$	(21), (10)
23	$K = \mathsf{FOr}$	(14), (15–22), (10)
24	$x \ \beta \ A$	(hyp.)
25	$\sim(x \ \beta \ \overset{n}{\triangle G})$	(hyp.)
26	$(\exists z)(z = F^{:x}(n))$	(IT)
27	$d = F^{:x}(n)$	(26)
28	$\{xd\} \ \beta \ \overset{n}{G}$	(24), (26), (2)
29	$x \ \beta \ \overset{n}{\triangle G}$	(28)
30	$A \subseteq \overset{n}{\triangle G}$	(24–29)
31	$\overset{n}{\triangle G} = A$	(30), (2)
32	$(x)(x \ \beta \ A \supset \overset{n}{G}(x) = F^{:x}(n))$	(31), (2)

Thus, if we take FOr for A, thereby making F a function which assigns to every finite ordinal a finite ordinal, we have, by the existence theorem above (ET), that there exists the function $\overset{n}{F}$ for every finite ordinal n and every such function F (cf. § 4.3, p. 82).

4.5. Addition, Multiplication, and Exponentiation

We are now in a position to define the simplest of the arithmetic functions. This is done in terms of the function $\overset{n}{F}$ (with suitably selected F) by reference to the value of $\overset{n}{F}$ for some particular finite ordinal.

Definitions

Df.7. $n + k$ $=df$ $\overset{k}{F}(n)$ where $[(x)(F(x) = x') \cdot {}^{\triangle}F = \mathsf{FOr}]$

Df.8. $n \times k$ $=df$ $\overset{k}{F}(0)$ where $[(x)(F(x) = x + n) \cdot {}^{\triangle}F = \mathsf{FOr}]$

Df.9. k^n $=df$ $\overset{n}{F}(1)$ where $[(x)(F(x) = x \times k) \cdot {}^{\triangle}F = \mathsf{FOr}]$

With the help of these definitions we have the following equations as theorems of the system:

T41. $n + 0 = n$

Proof

1      $n + 0 = \overset{0}{F}(n) \cdot (x)(F(x) = x') \cdot {}^{\triangle}F = \mathsf{FOr}$
                                                    (Df. 7, p. 61)

2      $\{n, n + 0\} \beta \overset{0}{F}$ (1)

3      $n + 0 = F^{:n}(0)$ (ET)

4      $n + 0 = n$ (IT)

T42. $n \times 0 = 0$

Proof

(Analogous to the proof of T41)

T43. $n^0 = 1$

Proof

(Analogous to the proof of T41)

T44. $k + n' = (k + n)'$

Proof

1	$\overset{n'}{F}(k) = g \cdot \overset{n}{F}(k) = h$	(hyp.)
2	$(x)(F(x) = x')$	((1), Df. 7, p. 89)
3	$\{kg\} \beta \overset{n'}{F} \cdot \{kh\} \beta \overset{n}{F}$	(1)
4	$g = F^{:k}(n') \cdot h = F^{:k}(n)$	(ET)
5	$g = F(F^{:k}(n))$	(IT)
6	$F(F^{:k}(n)) = (F^{:k}(n))'$	(2)
7	$g = h'$	(4–6)
8	$k + n' = (k + n)'$	(1), (7)

T45. $k \times n' = (k \times n) + k$

Proof

(Analogous to the proof of T44)

T46. $k^{n'} = k^n \times k$

Proof

(Analogous to the proof of T44)

From T41–T46 the general laws governing addition, multiplication and exponentiation can be derived as theorems of the system by complete induction (T34).

THE THEORY OF FINITE CLASSES AND FINITE SETS

In this section we aim to show only the main lines of connection between the theory of finite ordinals and the theory of finite classes and finite sets. The most important point of connection is the use of *finite ordinal number* to define *number of a class*. For, without such a definition we cannot go on to prove any assertion (such as, that the number of sub-sets of a class having n members in 2^n,) which refers to the number of a class. But the theory of finite classes and sets is concerned with just such assertions. We shall, therefore, introduce the required definition and derive some of its immediate consequences, thus showing that the theory of finite classes and finite sets is available as an extension of the present system. This will conclude our exposition.

5.1. Preliminary Theorems on Finite Ordinals

T51. $(n)\{n\ \beta\ \mathsf{FOr} \supset (\mathrm{Fnc}\colon F \cdot n \leftrightarrow {}^{\triangle}F \supset (\exists y)(y \leftrightarrow {}^{\triangle\smile}F))\}$

(If the domain of a function is represented by a finite ordinal, then the converse domain is represented by a set.)

Proof

1	$\vdash^{\smallfrown}[\mathrm{Fnc}\colon F \cdot n \leftrightarrow {}^{\triangle}F \supset (\exists y)(y \leftrightarrow {}^{\triangle\smile}F)]_F^n$	(CT)
2	Let the class so defined be C.	
3	$\mathrm{Fnc}\colon F \cdot 0 \leftrightarrow {}^{\triangle}F$	(hyp.)
4	${}^{\triangle}F \neq \wedge \supset (\exists z)(z\ \varepsilon\ 0)$	((3), T6, p. 49)
5	$F \neq \wedge \supset {}^{\triangle}F \neq \wedge$	(AxIIIc(i), p. 45)
6	${}^{\triangle\smile}F \neq \wedge \supset F \neq \wedge$	(AxIIIc(i)–(ii), p. 45)
7	${}^{\triangle\smile}F \neq \wedge \cdot 0 \leftrightarrow \wedge$	((4–6), T1 and T6, p. 49)

91

8    $(\exists y)(y \leftrightarrow {}^{\vartriangle\smile}F)$ (7)

9    $0 \, \beta \, C$ (1), (3–8)

10   $k \, \beta \, C$ (hyp.)

11   $(\text{Fnc}: F \cdot k \leftrightarrow {}^{\vartriangle}F) \supset (\exists y)(y \leftrightarrow {}^{\vartriangle\smile}F)$
                                   (1), (10)

12   $\text{Fnc}: G \cdot k' \leftrightarrow {}^{\vartriangle}G$ (hyp.)

13   $\vdash^{\widehat{}}[\text{Fnc}: F \cdot n' \leftrightarrow {}^{\vartriangle}F \cdot z \, \beta \, /\text{mem}_1 n \, \cap \, F/]^z_{F,n}$
                                   (CT)

14   Let the class so defined be $B_{F,n}$.

15   $B_{G,k} \subseteq G$ (13)

16   $\text{Fnc}: B_{G,k}$ ((12), (15), Df. 1, p. 65)

17   $k \leftrightarrow {}^{\vartriangle}B_{G,k}$ (13)

18   $d \leftrightarrow {}^{\vartriangle\smile}B_{G,k}$ (11), (16), (17)

19   $s^{*d,G(k)} \leftrightarrow {}^{\vartriangle\smile}G$ ((15), (18), T2, p. 48)

20   $(\exists y)(y \leftrightarrow {}^{\vartriangle\smile}G)$ (19)

21   $(\text{Fnc}: G \cdot k' \leftrightarrow {}^{\vartriangle}G) \supset (\exists y)(y \leftrightarrow {}^{\vartriangle\smile}G)$
                                   (12–20)

22   $k' \, \beta \, C$ (1), (21)

23   $(n)(n \, \beta \, \mathsf{FOr} \supset n \, \beta \, C)$ ((9), (10–22), T34, p. 81)

T52. $(A \subseteq n \cdot n \, \beta \, \mathsf{FOr}) \supset (\exists y)(y \leftrightarrow A)$

(Every sub-class of a finite ordinal is represented by a set.)

Proof

1    $A \subseteq n \cdot n \, \beta \, \mathsf{FOr}$ (hyp.)

2    $\text{vct}: A \supset 0 \leftrightarrow A$ (T1, p. 48)

3    $\sim\!\text{vct}: A \cdot p \, \beta \, A$ (hyp.)

4    $B = /\text{ldPr} \, \cap \, \text{mem}_1 A/$ (hyp.)

5    $\vdash^{\widehat{}}[c \, \varepsilon \, n \cdot \sim(c \, \beta \, A) \cdot d = p]^{cd}$ (CT)

6   Let D be the class so defined.

7    $x,y \, \beta \, /B \cup D/ \cdot x \neq y \cdot x = \{qr\} \cdot y = \{fg\}$
                                   (hyp.)

8    $q = f$ (hyp.)

9    $x,y \, \beta \, B \supset r = g$ (4), (7), (8)

10    $x,y \, \beta \, D \supset r = g$ (5–8)

11    $(x \, \beta \, B \cdot y \, \beta \, D) \supset (f \, \beta \, A \cdot \sim(f \, \beta \, A))$ (4–8)

12   Fnc: $|B \cup D| \cdot n \leftrightarrow {}^{\triangle}|B \cup D| \cdot {}^{\triangle\vee}|B \cup D| = A$
                              ((7–11), Df. 1, p. 65)

13    $(\exists y)(y \leftrightarrow A)$ (T51, p. 91)

Df.10. $H \approx \dfrac{A}{n}$ $=df$ (DbFnc: $H \cdot A = {}^{\triangle}H \cdot n \leftrightarrow {}^{\triangle\vee}H$)

T53. $(n \, \beta \, \mathsf{FOr} \cdot H \approx \dfrac{A}{n}) \supset (\exists yz)(y \leftrightarrow A \cdot z \leftrightarrow H)$

(If there is a one-to-one correspondence between a class A and a finite ordinal, then A and the one-to-one correspondence are each represented by a set.)

Proof

1    $H \approx \dfrac{A}{n} \cdot n \, \beta \, \mathsf{FOr}$ (hyp.)

2   Fnc: ${}^{\vee}H$ (Df. 3, p. 65)

3    $b \leftrightarrow {}^{\triangle\vee\vee}H$ (T51, Df. 10)

4    $b \leftrightarrow A$ ((3), (1), Df. 10)

5    $G = \left/\!\!\left/\left|{}^{\triangle\vee}_{\mathstrut}\begin{matrix}\mathrm{mem}_1 H \\ \mathrm{mem}_1 H\end{matrix}\mathsf{Pr} \cap \mathsf{IdPr}\right|\overrightarrow{}\right/\right.$ (hyp.)

6    $x,y \, \beta \, G \cdot x \neq y \cdot x = \{a\{ba\}\} \cdot y = \{c\{dc\}\}$
                              (hyp.)

7    $\{ba\},\{dc\} \, \beta \, H \cdot \mathrm{DbFnc}: H$ (5), (6)

8    $a = c \supset b = d$ (7)

9   Fnc: G (6–8)

10    ${}^{\triangle}G = {}^{\triangle\vee}H$ (5)

11    $n \leftrightarrow {}^{\triangle}G$ (1), (5)

12    $(\exists y)(y \leftrightarrow {}^{\triangle\vee}G) \cdot H = {}^{\triangle\vee}G$ (T51, p. 91)

13    $(\exists yz)(y \leftrightarrow A \cdot z \leftrightarrow H)$ (4), (12)

T54. $(z)(z \, \beta \, \approx\mathsf{Pr} \equiv (\exists x,y)(z = \{xy\} \cdot x \, \beta \, \mathsf{FOr} \cdot x \approx y))$

(There exists the class of those pairs $\{xy\}$ such that x is a finite ordinal and is one-to-one with y.)

Proof

1    $\vdash^\frown [x \, \beta \, \text{FOr} \cdot H \approx^x_y]^{xy}_H$ (CT)

2   Let the class so defined be $\approx$Pr.

3    $z \, \beta \approx\text{Pr} \supset (\exists xy)(z = \{xy\} \cdot x \, \beta \, \text{FOr} \cdot x \approx y)$
                                                    (Df. 10, p. 93)

4    $x \approx y \cdot x \, \beta \, \text{FOr} \cdot z = \{xy\}$ (hyp.)

5    $(\exists H)H \approx^x_y$ (Df. 4, p. 65) and
                                                    (Df. 10, p. 93)

6    $z \, \beta \approx\text{Pr}$ (1), (4), (5)

7    $(z)(z \, \beta \approx\text{Pr} \equiv (\exists x,y)(z = \{xy\} \cdot x \, \beta \, \text{FOr} \cdot x \approx y))$
                                                    (3), (4–7)

T55. $A \approx A$

(Every class is one-to-one with itself.)

Proof

1    $C = \cap(\text{mem}_1 A, \, \text{mem}_2 A)$ (hyp.)

2    $D = \cap(\text{IdPr}, \, C)$ (hyp.)

3    $\{ab\},\{cd\} \, \beta \, D \cdot \{ab\} \neq \{cd\}$ (hyp.)

4    $a = c \supset b = d$ (2), (3)

5   Fnc: D (3), (4)

6    $\{ab\},\{cd\} \, \beta \, {}^\smile D \cdot \{ab\} \neq \{cd\}$ (hyp.)

7    $a = c$ (hyp.)

8    $\{ba\},\{dc\} \, \beta \, D \cdot b = a \cdot d = c$ (6), (2)

9    $b = d$ (7), (8)

10   DbFnc: $D \cdot D \approx^A_A$ ((5), (6–9), Df. 10,
                                                    p. 93)

11   $A \approx A$ ((10), Df. 4, p. 65)

T56. $(A \approx B) \supset (B \approx A)$

(If a class is one-to-one with another class, then the latter is one-to-one
with the former.)

Proof

1    $H \approx \dfrac{A}{B}$ (hyp.)

2   DbFnc: $\smile H \cdot H = \smile H$ (1)

3    $\smile H \approx \dfrac{B}{A}$ (1), (2)

4    $B \approx A$ (3)

T57. $(A \approx B \cdot B \approx C) \supset A \approx C$

(If one class is one-to-one with another which is one-to-one with a third, then the first is one-to-one with the third.)

Proof

1    $A \approx B \cdot B \approx C$ (hyp.)

2    $G \approx \dfrac{A}{B} \cdot H \approx \dfrac{B}{C}$ (1)

3    $\{dr\},\{fg\} \,\beta\, \mathrm{Cmp} \dfrac{G}{H} \cdot \{dr\} \neq \{fg\}$ (hyp.)

4    $d = f$ (hyp.)

5    $\{dq\},\{fp\} \,\beta\, G \cdot \{qr\},\{pg\} \,\beta\, H$ ((3), Cmp. Lemma, p. 59)

6   DbFnc: $G \cdot$ DbFnc: H (2)

7    $p = q \cdot r = g$ (5), (6)

8    $d \neq f$ (3), (4–7)

9    $r \neq g$ (sim.)

10   DbFnc: $\mathrm{Cmp} \dfrac{G}{H} \cdot {}^{\triangle}\mathrm{Cmp} \dfrac{G}{H} = A$ ((2–9), Cmp. Lemma)

        $\cdot\, {}^{\triangle\smile}\mathrm{Cmp} \dfrac{G}{H} = C$

11   $A \approx C$ ((10), Df. 4, p. 65)

Note: Where a and b are sets representing classes A and B respectively, we shall use $a \approx b$, $a \approx B$, $A \approx b$, in place of $A \approx B$.

T58. $(n \,\beta\, \mathrm{FOr} \cdot s \subset n) \supset (\exists\mathrm{W})(\exists z)(z \,\beta\, \mathrm{L}_n \cdot \mathrm{W} \approx \dfrac{z}{s})$

(For every proper sub-set s of a finite ordinal there exists a one-to-one correspondence between s and an ordinal lower than n.)

Proof

1	$n\ \beta\ \mathsf{FOr} \cdot s \subset n$	(hyp.)
2	$\vdash^{\frown}[n\ \beta\ \mathsf{FOr} \cdot (x)(x \subset n \supset (\exists y)(y \subset n \cdot \{yx\}\ \beta \approx \mathsf{Pr}]^n$	(CT)
3	Let A be the class so defined.	
4	$0\ \beta\ A$	(2)
5	$k\ \beta\ A$	(hyp.)
6	$r \subset k'$	(hyp.)
7	$r = k \supset k'\ \beta\ A$	((2), T54, p. 93)
8	$r \subset k \supset k'\ \beta\ A$	((2), T54, p. 93)
9	$\sim(r \subseteq k)$	(hyp.)
10	$a\ \varepsilon\ r \cdot \sim(a\ \varepsilon\ k)$	(9)
11	$(a = k)$	(6)
12	$k\ \varepsilon\ r$	(10), (11)
13	$\vdash^{\frown}[b\ \varepsilon\ r \cdot b \neq k]^b$	(CT)
14	Let P be the class so defined.	
15	$g \leftrightarrow P$	(T52, p. 92)
16	$b\ \varepsilon\ k' \cdot \sim(b\ \varepsilon\ r)$	(6)
17	$b \neq k$	((12), (16), FAx, p. 45)
18	$b\ \varepsilon\ k$	(16), (17)
19	$(\exists x)(x\ \varepsilon\ k \cdot \sim(x\ \varepsilon\ r))$	(16), (18)
20	$P \subseteq k$	(13), (6)
21	$(\exists x)(x\ \varepsilon\ k \cdot \sim(x\ \beta\ P))$	(13), (19)
22	$P \subset k \cdot g \subset k \cdot k\ \beta\ A$	(20), (21)
23	$(\exists y)(y \subset k \cdot \{yg\}\ \beta \approx \mathsf{Pr})$	(2), (22)
24	$H \approx^g_y \cdot g \leftrightarrow {}^{\triangle}H \cdot y \leftrightarrow {}^{\triangle\smallsmile}H$	(T56, T54, p. Df. 10, p. 93)
25	$\sim(k\ \varepsilon\ g) \cdot \sim(y\ \varepsilon\ y)$	((13), (15), T17, p. 64)
26	$\mathrm{DbFnc}\colon |H \cup (\{ky\})| \cdot r \overset{\triangle}{\leftrightarrow} |H \cup (\{ky\})|$ $\cdot y' \overset{\triangle\smallsmile}{\leftrightarrow} |H \cup (\{ky\})|$	(24), (25)

27	$/H \cup (\{ky\})/ \approx^{r}_{y'} \cdot y' \, \beta \, \mathsf{FOr}$	(26), (23)
28	$y \subset k \cdot y \, \varepsilon \, k$	(T22.2, p. 71)
29	$y' \subseteq k' \cdot {\sim}(k \, \varepsilon \, y')$	(28)
30	$y' \subset k' \cdot \{y'r\} \, \beta \approx\mathsf{Pr}$	((27), (29), T54, p. 93)
31	$r \subseteq k \supset (r \subset k \lor r = k)$	(dfs., p. 44)
32	$(r \subset k \lor r = k) \supset (k' \, \beta \, A)$	(7), (8)
33	${\sim}(r \subseteq k) \supset (\exists y)(y \subset k' \cdot \{yr\} \, \beta \approx\mathsf{Pr})$	(9–30)
34	${\sim}(r \, \varepsilon \, k) \supset k' \, \beta \, A$	((31–33), T22.2, p. 71)
35	$\mathsf{FOr} \subseteq A$	(T34, p. 81)
36	$(\exists y)(y \subset n \cdot \{ys\} \, \beta \approx\mathsf{Pr})$	(1), (3), (35)
37	$y \, \beta \, \mathsf{FOr} \cdot G \approx^{y}_{s}$	(36)
38	$(\exists F)(\exists z)(z \, \beta \, \mathrm{L}_n \cdot F \approx^{z}_{s})$	((36), (37), Df.C, p. 75)

T59. $(k \, \beta \, \mathsf{Or} \cdot n \, \beta \, \mathsf{FOr} \cdot k \approx n) \supset (k = n)$

(There cannot be a one-to-one correspondence between a finite ordinal n and an ordinal distinct from n.)

Proof

1	$n \, \beta \, \mathsf{FOr} \cdot k \approx n \cdot k \, \beta \, \mathsf{Or}$	(hyp.)
2	$\vdash\!\hat{\ }[n \, \beta \, \mathsf{FOr} \cdot (\exists y)(\{yn\} \, \beta \approx\mathsf{Pr} \cdot y \neq n \cdot y \, \beta \, \mathsf{Or})]^n$	(CT)
3	Let A be the class so defined.	
4	${\sim}\mathrm{vct}\!: A$	(hyp.)
5	$b \, \beta \, A \cdot (z)(z \, \beta \, A \cdot z \neq b \supset b \subset z)$	(T24.4, p. 74)
6	$\{yb\} \, \beta \approx\mathsf{Pr} \cdot y \neq b \cdot H \approx^{b}_{y}$	(5), (3)
7	$y \, \beta \, \mathsf{Or} \cdot \{by\} \, \beta \approx\mathsf{Pr} \cdot b \neq y$	((2), T56, p. 94)
8	$y \, \beta \, A \cdot y \neq b$	(3), (7)
9	$b \subset y \cdot b \approx y \cdot b \leftrightarrow {}^{\triangle}H \cdot y \leftrightarrow {}^{\triangle\smallsmile}H$	(8), (5), (3)
10	$c \subset y$	(hyp.)

11	$D = /H \cap \mathrm{mem}_2 c/$	(CT)
12	$\mathrm{DbFnc}\colon H \cdot D \subseteq H$	(6), (11)
13	$(x)(x\,\beta\,D \supset (\exists y)(x = \{yb\} \cdot b\,\varepsilon\,c))$	(11)
14	$(D = H) \supset (c \leftrightarrow {}^{\vartriangle\smallsmile}H)$	(11)
15	$c \subset {}^{\vartriangle\smallsmile}H$	(9), (10)
16	$D \subset H \cdot \mathrm{DbFnc}\colon D$	(12), (14), (15)
17	${}^{\vartriangle}D \subset {}^{\vartriangle}H$	(16)
18	${}^{\vartriangle}D \subset b$	(9), (17)
19	$g \leftrightarrow {}^{\vartriangle}D \cdot g \leftrightarrow P$	((18), T52, p. 92; T15, p. 51)
20	$c \leftrightarrow {}^{\vartriangle\smallsmile}D$	(11)
21	$D \approx_{c}^{P} \cdot P \subset b$	((Df. 10, p. 93, (18), (19))
22	$(\exists W)(W \subset b \cdot W \approx c)$	(21)
23	$(x)(x \subset y \supset (\exists W)(W \subset b \cdot W \approx x))$	(10–22)
24	$W \subset b \cdot W \approx b$	(9), (23)
25	$s \leftrightarrow W \cdot s \subset b \cdot s \approx b$	(T52, p. 92)
26	$(\exists z)(\{sz\}\,\beta \approx\!\mathrm{Pr} \cdot z \subset b \cdot z\,\beta\,\mathrm{Or})$	(T58, p. 95)
27	$z\,\beta\,\mathrm{FOr}$	(T30.2, p. 81)
28	$s \approx z \cdot z \subset b$	(26)
29	$z \approx b \cdot {\sim}(z = b)$	(T57, p. 95)
30	$z\,\beta\,A \cdot b \subset z$	(27), (29), (3), (5)
31	$\mathrm{vct}\colon A \cdot \{kn\}\,\beta \approx\!\mathrm{Pr}$	(4–28), (29), (30), (1)
32	$k = n$	(30), (31), (3)

Df.E. FSt *for* the class determined by: $[a\,\beta\,{}^{\vartriangle\smallsmile}\!\approx\!\mathrm{Pr}]^a$

(A class or set will be called **finite** if it is one-to-one with a finite ordinal.)

By the next theorem we show that the class of finite ordinals is the class of finite sets which are also ordinals. We conclude with a group of five theorems on finite classes and finite sets, the third of which permits the introduction of *the number of a class.* The last two theorems of the group make essential use of this concept.

5.2. Theorems on Finite Classes and Sets

T60. $|\mathsf{FSt} \cap \mathsf{Or}| = \mathsf{FOr}$

(The class of finite ordinals is the intersection of the class of finite sets and the class of ordinals.)

Proof

1	$a \, \beta \, |\mathsf{FSt} \cap \mathsf{Or}|$	(hyp.)
2	$a \, \beta \, \mathsf{Or} \cdot b \, \beta \, \mathsf{FOr} \cdot a \approx b$	((1), Df.E, p. 98)
3	$\sim(a \, \beta \, \mathsf{FOr})$	(hyp.)
4	$a \approx b \cdot a \neq b \cdot a \, \beta \, \mathsf{Or}$	((2), (3), FAx, p. 45)
5	$a \, \beta \, \mathsf{FOr}$	((4), T59, p. 97)
6	$(a \, \beta \, \mathsf{FOr} \cdot a \approx a) \supset (a \, \beta \, |\mathsf{FSt} \cap \mathsf{Or}|)$	(Df.D, p. 55; Df.E, p. 98)
7	$|\mathsf{FSt} \cap \mathsf{Or}| = \mathsf{FOr}$	((1–5), (6), T55, p. 94)

Df.11. $\mathsf{FCl}: C =_{df} (\exists x)(x \, \beta \, \mathsf{FOr} \cdot x \approx C)$

(C is a finite class.)

T61. $\mathsf{FCl}: A \supset (\exists x)(x \, \beta \, \mathsf{FSt} \cdot x \leftrightarrow A)$

(Every finite class is represented by a finite set.)

Proof

1	$\mathsf{FCl}: A$	(hyp.)
2	$b \, \beta \, \mathsf{FOr} \cdot b \approx A$	((1), df. 11)
3	$(\exists x)(x \leftrightarrow A)$	(T53, p. 93)
4	$c \leftrightarrow A \cdot b \approx c$	((2), (3), Df. 4, p. 65)
5	$c \, \beta \, \mathsf{FSt}$	((2), (4), Df.E, p. 98)
6	$(\exists x)(x \, \beta \, \mathsf{FSt} \cdot x \leftrightarrow A)$	(4), (5)

T62. $(C \subseteq D \cdot \mathsf{FCl}: D) \supset \mathsf{FCl}: C$

(Every sub-class of a finite class is finite.)

Proof

1	FCl: $D \cdot C \subseteq D$	(hyp.)
2	$k \, \beta \, \mathsf{FOr} \cdot D \approx k$	((1), Df. 11)
3	$(C = D) \supset (\mathrm{FCl}\colon C)$	(1)
4	$C \subset D$	(hyp.)
5	$k \leftrightarrow A \cdot H \underset{D}{\approx} \frac{A}{} \cdot {}^{\vartriangle}H = A \cdot {}^{\vartriangle \smallsmile}H = D$	
		((2), T15, p. 51)
6	$/H \cap \mathrm{mem}_2 C/ \subseteq H$	(AxIII(a3), p. 45)
7	$/H \cap \mathrm{mem}_2 C/ \underset{C}{\approx} \frac{P}{} \cdot P \subset A \cdot C \subset D \cdot P \subset k \cdot g \leftrightarrow P$	
		(T52, p. 92)
8	$g \subset k \cdot (\exists x)(x \, \beta \, \mathrm{L}_k \cdot x \approx g) \cdot C \approx g$	((7), T58, p. 95)
9	$n \, \beta \, \mathsf{Or} \cdot n \approx g \cdot n \approx C \cdot n \subset k$	((8), Df.C, p. 75)
10	$n \, \beta \, \mathsf{FOr}$	((9), T30.2, p. 81)
11	FCl: C	((9), (10), Df. 11, p. 99)

T63. $(\mathrm{FCl}\colon D \supset (\exists \, ! \, x)(x \, \beta \, \mathsf{FOr} \cdot x \approx D))$
$\qquad\qquad \cdot (b \, \beta \, \mathsf{FSt} \supset (\exists \, ! \, x)(x \, \beta \, \mathsf{FOr} \cdot x \approx b))$

(There is only one finite ordinal which is one-to-one with a given finite class (or finite set).)

Proof
(Immediate from T59, Df.E and Df. 11.)

Note: **The number of the class** C (or **the number of the set** c) shall mean the unique finite ordinal which, by T63, is one-to-one with the class C (or the set c).

Thus, by RIC (cf. II.0.22, p. 40):

Df.F. $\begin{cases} z = N`C \; \equiv \; (z \, \beta \, \mathsf{FOr} \cdot z \approx C) \\ z = N`c \; \equiv \; (z \, \beta \, \mathsf{FOr} \cdot z \approx c) \end{cases}$

T64. $\{ \mathrm{FCl}\colon D \cdot n = N`D$
$\qquad\quad \cdot [(C \subset D \cdot k = N`C) \vee (b \subset D \cdot k = N`b)] \supset k \subset n \}$
$\quad \cdot \{ a \, \beta \, \mathsf{FSt} \cdot n = N`a$
$\qquad\quad \cdot [(C \subset a \cdot k = N`C) \vee (b \subset a \cdot k = N`b)] \supset k \subset n \}$

(The number of a proper sub-class (sub-set) of a finite class (set) is lower than the number of that class (set).)

Proof

In each of the four cases the proof is based on T58 and T63, as follows:

1	FCl: $D \cdot n = N'D \cdot b \subset D \cdot k = N'b$	(hyp.)
2	$H \approx \underset{n}{D} \cdot b \, \beta \, \mathsf{FSt}$	(Df.F and Df.E, p. 98)
3	$/H \cap \mathrm{mem}_1 b/ \approx \underset{P}{b} \cdot P \subset n$	(cf. T59; 12–22, p. 98)
4	$g \leftrightarrow P \cdot g \subset n \cdot b \approx g$	(T52, p. 92)
5	$a \subset n \cdot a \approx g \cdot a \, \beta \, \mathsf{Or}$	(T58, p. 95)
6	$a \, \beta \, \mathsf{FOr} \cdot b \approx a$	(T30.2, p. 81; T57, p. 95)
7	$a \approx k$	((1), (6), T57, p. 95)
8	$k \subset n$	(T59, p. 97)

T65. $(FCl: A \cdot B \subset A) \supset {\sim}(A \approx B)$

(There is no one-to-one correspondence between a finite class and a proper sub-class of it.)

Proof

1	FCl: $A \cdot B \subset A$	(hyp.)
2	$n = N'A \cdot C \approx \underset{n}{A}$	(T63)
3	$/C \cap \mathrm{mem}_1 B/ \approx \underset{P}{B} \cdot P \subset n \cdot s \leftrightarrow P$	(T52, p. 92)
4	$s \subset n \cdot s \approx k \cdot k \, \beta \, \mathsf{Or} \cdot k \subset n$	(T58, p. 95)
5	$A \approx B$	(hyp.)
6	$n \approx A \cdot B \approx s \cdot s \approx k$	(2), (3), (4)
7	$n \approx k \cdot k \subset n$	(T57, p. 95)
8	$n \approx k \cdot k \, \beta \, \mathsf{Or} \cdot k \neq n$	(4), (7)
9	${\sim}(A \approx B)$	((8), T59, p. 97)

5.3 Conclusion

In this system we have followed a course of precise argument leading from elementary truths about classes and sets, to elementary truths about numbers. In short we have drawn a continuous line between the logic of classes and the beginning of mathematics. The same line can be produced through the entire field of mathematics, although in doing so we need to add a few further axioms to the system (cf. the later parts of the Bernays system which appear in the *Journal of Symbolic Logic*, volumes 7, 8 and 13). That we should be able to do this is a matter of some interest from both a philosophical and a mathematical point of view. For, in the one case, the claim that mathematical truths can be shown, in a precise if indirect way, to be the offspring of logical truths, is thereby substantiated. In the other case, the mathematician is thereby absolved from taking refuge in the mystic belief: "God created the integers, the rest is the work of man".[1] But the pursuit of these lines of thought is beyond the scope of the present text.

[1] Kronecker, *Mathematische Annalen*, Vol. 43, 1893, p. 15.

REFERENCES

BASSON, A. H. and O'CONNOR, D. J., *Introduction to Symbolic Logic*, 2nd Ed., London, University Tutorial Press, 1957.

BERNAYS, PAUL, A System of Axiomatic Set Theory, *Journal of Symbolic Logic*, Vol. 2 (1937), pp. 65–77; Vol. 6 (1941), pp. 1–17; Vol. 7 (1942), pp. 65–89 and 133–145; Vol. 8 (1943), pp. 89–106; Vol. 13 (1948), pp. 65–79.

BERNAYS, PAUL and FRAENKEL, A. A., *Axiomatic Set Theory*, Amsterdam, North-Holland, 1958.

FRAENKEL, A. A., and BAR-HILLEL, Y., *Foundations of Set Theory*, Amsterdam, North-Holland, 1958.

HALMOS, P. R., *Naive Set Theory*, Princeton, Van Nostrand, 1960.

HEMPEL, CARL G., On the Nature of Mathematical Truth. *In:* FEIGL, H. and SELLARS, W., *Readings in Philosophical Analysis*, New York, Appleton-Century, 1949; pp. 222–237.

KERSHNER, R. B. and WILCOX, L. R., *The Anatomy of Mathematics*, New York, Ronald Press, 1950.

LANGER, SUSANNE K., *An Introduction to Symbolic Logic* (2nd Revised Ed.), New York, Dover Press, 1953.

QUINE, W. VAN ORMAN, *Mathematical Logic* (Revised Ed.), Cambridge, Harvard University Press, 1951.

ROSSER, J. B., *Logic for Mathematicians*, New York, McGraw-Hill, 1953.

RUSSELL, BERTRAND, *Introduction to Mathematical Philosophy*, London, Allen & Unwin, 1948.

SUPPES, PATRICK, *Introduction to Logic*, New York, Van Nostrand, 1957.

SUPPES, P., *Axiomatic Set Theory*, Princeton, Van Nostrand, 1960.

TARSKI, ALFRED, *Introduction to Logic*, New York, O.U.P., 1951.

WILDER, RAYMOND, L., *Introduction to the Foundations of Mathematics*, New York, Wiley, 1952.

INDEX

105